This book belongs to

	COLONY NAME	
	DATE	
	TIME	

WEATHER CONDITIONS

🌡 _____ ☀ ⛅ 🌧 ⛈ ❄

🏳 _____ ☐ ☐ ☐ ☐ ☐

INSPECTION

HIVE NUMBER	①	②	③	④	⑤	⑥

PRODUCTIVITY & REPRODUCTION

AMOUNT OF HONEY						
GENERAL POPULATION						
AMOUNT OF BROOD						
AMOUNT OF SPACE						

BEHAVIOUR & ACTIVITIES

USUAL ENTERING AND EXITING ACTIVITY?						
CALM BEHAVIOUR WHEN OPENING HIVE?						
BEES BRINGING POLLEN INTO HIVE?						
SIGNS OF ROBBERY AMONG THE BEES?						

HEALTH STATUS

BEES SEEM WEAK OR LAZY?						
HIGH AMOUNT OF DEAD BEES?						
QUEEN BEE IS PRESENT / IDENTIFIABLE?						
INFESTATION BY ANTS / ANTS PRESENT?						
INFESTATION BY WAX MOTH / WAX MOTH PRESENT?						
NEGATIVE ODOUR NOTICEABLE?						

COLONY NAME	WEATHER CONDITIONS

COLONY NAME
DATE
TIME

WEATHER CONDITIONS

INSPECTION

HIVE NUMBER	1	2	3	4	5	6

PRODUCTIVITY & REPRODUCTION

AMOUNT OF HONEY						
GENERAL POPULATION						
AMOUNT OF BROOD						
AMOUNT OF SPACE						

BEHAVIOUR & ACTIVITIES

USUAL ENTERING AND EXITING ACTIVITY?						
CALM BEHAVIOUR WHEN OPENING HIVE?						
BEES BRINGING POLLEN INTO HIVE?						
SIGNS OF ROBBERY AMONG THE BEES?						

HEALTH STATUS

BEES SEEM WEAK OR LAZY?						
HIGH AMOUNT OF DEAD BEES?						
QUEEN BEE IS PRESENT / IDENTIFIABLE?						
INFESTATION BY ANTS / ANTS PRESENT?						
INFESTATION BY WAX MOTH / WAX MOTH PRESENT?						
NEGATIVE ODOUR NOTICEABLE?						

COLONY NAME		WEATHER CONDITIONS					
DATE		🌡 ___	☀	⛅	☁	⛈	❄
TIME		🚩 ___	☐	☐	☐	☐	☐

INSPECTION

HIVE NUMBER	①	②	③	④	⑤	⑥

PRODUCTIVITY & REPRODUCTION

	1	2	3	4	5	6
AMOUNT OF HONEY						
GENERAL POPULATION						
AMOUNT OF BROOD						
AMOUNT OF SPACE						

BEHAVIOUR & ACTIVITIES

	1	2	3	4	5	6
USUAL ENTERING AND EXITING ACTIVITY?						
CALM BEHAVIOUR WHEN OPENING HIVE?						
BEES BRINGING POLLEN INTO HIVE?						
SIGNS OF ROBBERY AMONG THE BEES?						

HEALTH STATUS

	1	2	3	4	5	6
BEES SEEM WEAK OR LAZY?						
HIGH AMOUNT OF DEAD BEES?						
QUEEN BEE IS PRESENT / IDENTIFIABLE?						
INFESTATION BY ANTS / ANTS PRESENT?						
INFESTATION BY WAX MOTH / WAX MOTH PRESENT?						
NEGATIVE ODOUR NOTICEABLE?						

COLONY NAME	WEATHER CONDITIONS

COLONY NAME
DATE
TIME

WEATHER CONDITIONS

🌡 —— ☀ ⛅ 🌧 ⛈ ❄

🏳 —— ☐ ☐ ☐ ☐ ☐

INSPECTION

HIVE NUMBER	1	2	3	4	5	6

PRODUCTIVITY & REPRODUCTION

AMOUNT OF HONEY						
GENERAL POPULATION						
AMOUNT OF BROOD						
AMOUNT OF SPACE						

BEHAVIOUR & ACTIVITIES

USUAL ENTERING AND EXITING ACTIVITY?						
CALM BEHAVIOUR WHEN OPENING HIVE?						
BEES BRINGING POLLEN INTO HIVE?						
SIGNS OF ROBBERY AMONG THE BEES?						

HEALTH STATUS

BEES SEEM WEAK OR LAZY?						
HIGH AMOUNT OF DEAD BEES?						
QUEEN BEE IS PRESENT / IDENTIFIABLE?						
INFESTATION BY ANTS / ANTS PRESENT?						
INFESTATION BY WAX MOTH / WAX MOTH PRESENT?						
NEGATIVE ODOUR NOTICEABLE?						

COLONY NAME	WEATHER CONDITIONS

COLONY NAME

DATE

TIME

WEATHER CONDITIONS

🌡 — ☀ ⛅ 🌧 ⛈ ❄

🚩 — ☐ ☐ ☐ ☐ ☐

INSPECTION

HIVE NUMBER	1	2	3	4	5	6

PRODUCTIVITY & REPRODUCTION

AMOUNT OF HONEY						
GENERAL POPULATION						
AMOUNT OF BROOD						
AMOUNT OF SPACE						

BEHAVIOUR & ACTIVITIES

USUAL ENTERING AND EXITING ACTIVITY?						
CALM BEHAVIOUR WHEN OPENING HIVE?						
BEES BRINGING POLLEN INTO HIVE?						
SIGNS OF ROBBERY AMONG THE BEES?						

HEALTH STATUS

BEES SEEM WEAK OR LAZY?						
HIGH AMOUNT OF DEAD BEES?						
QUEEN BEE IS PRESENT / IDENTIFIABLE?						
INFESTATION BY ANTS / ANTS PRESENT?						
INFESTATION BY WAX MOTH / WAX MOTH PRESENT?						
NEGATIVE ODOUR NOTICEABLE?						

COLONY NAME		WEATHER CONDITIONS					
DATE		🌡 ____	☀	⛅	🌧	⛈	❄
TIME		🚩 ____	☐	☐	☐	☐	☐

INSPECTION

HIVE NUMBER	(1)	(2)	(3)	(4)	(5)	(6)

PRODUCTIVITY & REPRODUCTION

	1	2	3	4	5	6
AMOUNT OF HONEY						
GENERAL POPULATION						
AMOUNT OF BROOD						
AMOUNT OF SPACE						

BEHAVIOUR & ACTIVITIES

	1	2	3	4	5	6
USUAL ENTERING AND EXITING ACTIVITY?						
CALM BEHAVIOUR WHEN OPENING HIVE?						
BEES BRINGING POLLEN INTO HIVE?						
SIGNS OF ROBBERY AMONG THE BEES?						

HEALTH STATUS

	1	2	3	4	5	6
BEES SEEM WEAK OR LAZY?						
HIGH AMOUNT OF DEAD BEES?						
QUEEN BEE IS PRESENT / IDENTIFIABLE?						
INFESTATION BY ANTS / ANTS PRESENT?						
INFESTATION BY WAX MOTH / WAX MOTH PRESENT?						
NEGATIVE ODOUR NOTICEABLE?						

COLONY NAME		WEATHER CONDITIONS					
DATE		🌡 ___	☀	⛅	🌧	⛈	❄
TIME		🚩 ___	☐	☐	☐	☐	☐

INSPECTION

HIVE NUMBER	①	②	③	④	⑤	⑥

PRODUCTIVITY & REPRODUCTION

AMOUNT OF HONEY						
GENERAL POPULATION						
AMOUNT OF BROOD						
AMOUNT OF SPACE						

BEHAVIOUR & ACTIVITIES

USUAL ENTERING AND EXITING ACTIVITY?						
CALM BEHAVIOUR WHEN OPENING HIVE?						
BEES BRINGING POLLEN INTO HIVE?						
SIGNS OF ROBBERY AMONG THE BEES?						

HEALTH STATUS

BEES SEEM WEAK OR LAZY?						
HIGH AMOUNT OF DEAD BEES?						
QUEEN BEE IS PRESENT / IDENTIFIABLE?						
INFESTATION BY ANTS / ANTS PRESENT?						
INFESTATION BY WAX MOTH / WAX MOTH PRESENT?						
NEGATIVE ODOUR NOTICEABLE?						

	COLONY NAME		WEATHER CONDITIONS

	COLONY NAME
	DATE
	TIME

WEATHER CONDITIONS

🌡 — ☀ ⛅ ☁ 🌧 ❄

🚩 — ☐ ☐ ☐ ☐ ☐

INSPECTION

	HIVE NUMBER	①	②	③	④	⑤	⑥

PRODUCTIVITY & REPRODUCTION

	AMOUNT OF HONEY						
	GENERAL POPULATION						
	AMOUNT OF BROOD						
	AMOUNT OF SPACE						

BEHAVIOUR & ACTIVITIES

	USUAL ENTERING AND EXITING ACTIVITY?						
	CALM BEHAVIOUR WHEN OPENING HIVE?						
	BEES BRINGING POLLEN INTO HIVE?						
	SIGNS OF ROBBERY AMONG THE BEES?						

HEALTH STATUS

	BEES SEEM WEAK OR LAZY?						
	HIGH AMOUNT OF DEAD BEES?						
	QUEEN BEE IS PRESENT / IDENTIFIABLE?						
	INFESTATION BY ANTS / ANTS PRESENT?						
	INFESTATION BY WAX MOTH / WAX MOTH PRESENT?						
	NEGATIVE ODOUR NOTICEABLE?						

COLONY NAME	
DATE	
TIME	

WEATHER CONDITIONS

🌡 —— ☀ ⛅ ☁ 🌧 ❄
🚩 —— ☐ ☐ ☐ ☐ ☐

INSPECTION

HIVE NUMBER	①	②	③	④	⑤	⑥

PRODUCTIVITY & REPRODUCTION

AMOUNT OF HONEY						
GENERAL POPULATION						
AMOUNT OF BROOD						
AMOUNT OF SPACE						

BEHAVIOUR & ACTIVITIES

USUAL ENTERING AND EXITING ACTIVITY?						
CALM BEHAVIOUR WHEN OPENING HIVE?						
BEES BRINGING POLLEN INTO HIVE?						
SIGNS OF ROBBERY AMONG THE BEES?						

HEALTH STATUS

BEES SEEM WEAK OR LAZY?						
HIGH AMOUNT OF DEAD BEES?						
QUEEN BEE IS PRESENT / IDENTIFIABLE?						
INFESTATION BY ANTS / ANTS PRESENT?						
INFESTATION BY WAX MOTH / WAX MOTH PRESENT?						
NEGATIVE ODOUR NOTICEABLE?						

COLONY NAME		WEATHER CONDITIONS					

🏠 COLONY NAME
📅 DATE
🕐 TIME

WEATHER CONDITIONS

🌡️ ——   ☀️ ⛅ 🌧️ ⛈️ ❄️

🚩 ——   ☐   ☐   ☐   ☐   ☐

INSPECTION

🍯 HIVE NUMBER	①	②	③	④	⑤	⑥

PRODUCTIVITY & REPRODUCTION

🍯 AMOUNT OF HONEY						
🐝 GENERAL POPULATION						
🐝 AMOUNT OF BROOD						
🍯 AMOUNT OF SPACE						

BEHAVIOUR & ACTIVITIES

🐝 USUAL ENTERING AND EXITING ACTIVITY?						
📦 CALM BEHAVIOUR WHEN OPENING HIVE?						
🐝 BEES BRINGING POLLEN INTO HIVE?						
🐝 SIGNS OF ROBBERY AMONG THE BEES?						

HEALTH STATUS

🐝 BEES SEEM WEAK OR LAZY?						
☠️ HIGH AMOUNT OF DEAD BEES?						
🐝 QUEEN BEE IS PRESENT / IDENTIFIABLE?						
🐜 INFESTATION BY ANTS / ANTS PRESENT?						
🦋 INFESTATION BY WAX MOTH / WAX MOTH PRESENT?						
👃 NEGATIVE ODOUR NOTICEABLE?						

COLONY NAME	WEATHER CONDITIONS

COLONY NAME	
DATE	
TIME	

INSPECTION

HIVE NUMBER	(1)	(2)	(3)	(4)	(5)	(6)

PRODUCTIVITY & REPRODUCTION

AMOUNT OF HONEY						
GENERAL POPULATION						
AMOUNT OF BROOD						
AMOUNT OF SPACE						

BEHAVIOUR & ACTIVITIES

USUAL ENTERING AND EXITING ACTIVITY?						
CALM BEHAVIOUR WHEN OPENING HIVE?						
BEES BRINGING POLLEN INTO HIVE?						
SIGNS OF ROBBERY AMONG THE BEES?						

HEALTH STATUS

BEES SEEM WEAK OR LAZY?						
HIGH AMOUNT OF DEAD BEES?						
QUEEN BEE IS PRESENT / IDENTIFIABLE?						
INFESTATION BY ANTS / ANTS PRESENT?						
INFESTATION BY WAX MOTH / WAX MOTH PRESENT?						
NEGATIVE ODOUR NOTICEABLE?						

	COLONY NAME		WEATHER CONDITIONS					
	DATE		🌡 ___	☀	⛅	🌧	⛈	❄
	TIME		🚩 ___	☐	☐	☐	☐	☐

INSPECTION

	HIVE NUMBER	(1)	(2)	(3)	(4)	(5)	(6)

PRODUCTIVITY & REPRODUCTION

	AMOUNT OF HONEY						
	GENERAL POPULATION						
	AMOUNT OF BROOD						
	AMOUNT OF SPACE						

BEHAVIOUR & ACTIVITIES

	USUAL ENTERING AND EXITING ACTIVITY?						
	CALM BEHAVIOUR WHEN OPENING HIVE?						
	BEES BRINGING POLLEN INTO HIVE?						
	SIGNS OF ROBBERY AMONG THE BEES?						

HEALTH STATUS

	BEES SEEM WEAK OR LAZY?						
	HIGH AMOUNT OF DEAD BEES?						
	QUEEN BEE IS PRESENT / IDENTIFIABLE?						
	INFESTATION BY ANTS / ANTS PRESENT?						
	INFESTATION BY WAX MOTH / WAX MOTH PRESENT?						
	NEGATIVE ODOUR NOTICEABLE?						

COLONY NAME	
DATE	
TIME	

INSPECTION

HIVE NUMBER	1	2	3	4	5	6

PRODUCTIVITY & REPRODUCTION

AMOUNT OF HONEY						
GENERAL POPULATION						
AMOUNT OF BROOD						
AMOUNT OF SPACE						

BEHAVIOUR & ACTIVITIES

USUAL ENTERING AND EXITING ACTIVITY?						
CALM BEHAVIOUR WHEN OPENING HIVE?						
BEES BRINGING POLLEN INTO HIVE?						
SIGNS OF ROBBERY AMONG THE BEES?						

HEALTH STATUS

BEES SEEM WEAK OR LAZY?						
HIGH AMOUNT OF DEAD BEES?						
QUEEN BEE IS PRESENT / IDENTIFIABLE?						
INFESTATION BY ANTS / ANTS PRESENT?						
INFESTATION BY WAX MOTH / WAX MOTH PRESENT?						
NEGATIVE ODOUR NOTICEABLE?						

COLONY NAME		WEATHER CONDITIONS					
DATE		🌡 ___	☀	⛅	☁	🌧	❄
TIME		🚩 ___	☐	☐	☐	☐	☐

INSPECTION

HIVE NUMBER	①	②	③	④	⑤	⑥

PRODUCTIVITY & REPRODUCTION

AMOUNT OF HONEY						
GENERAL POPULATION						
AMOUNT OF BROOD						
AMOUNT OF SPACE						

BEHAVIOUR & ACTIVITIES

USUAL ENTERING AND EXITING ACTIVITY?						
CALM BEHAVIOUR WHEN OPENING HIVE?						
BEES BRINGING POLLEN INTO HIVE?						
SIGNS OF ROBBERY AMONG THE BEES?						

HEALTH STATUS

BEES SEEM WEAK OR LAZY?						
HIGH AMOUNT OF DEAD BEES?						
QUEEN BEE IS PRESENT / IDENTIFIABLE?						
INFESTATION BY ANTS / ANTS PRESENT?						
INFESTATION BY WAX MOTH / WAX MOTH PRESENT?						
NEGATIVE ODOUR NOTICEABLE?						

COLONY NAME

DATE

TIME

WEATHER CONDITIONS

		☀	⛅	☁	🌧	❄
🌡	—					
	—	☐	☐	☐	☐	☐

INSPECTION

	1	2	3	4	5	6
HIVE NUMBER	①	②	③	④	⑤	⑥

PRODUCTIVITY & REPRODUCTION

AMOUNT OF HONEY						
GENERAL POPULATION						
AMOUNT OF BROOD						
AMOUNT OF SPACE						

BEHAVIOUR & ACTIVITIES

USUAL ENTERING AND EXITING ACTIVITY?						
CALM BEHAVIOUR WHEN OPENING HIVE?						
BEES BRINGING POLLEN INTO HIVE?						
SIGNS OF ROBBERY AMONG THE BEES?						

HEALTH STATUS

BEES SEEM WEAK OR LAZY?						
HIGH AMOUNT OF DEAD BEES?						
QUEEN BEE IS PRESENT / IDENTIFIABLE?						
INFESTATION BY ANTS / ANTS PRESENT?						
INFESTATION BY WAX MOTH / WAX MOTH PRESENT?						
NEGATIVE ODOUR NOTICEABLE?						

COLONY NAME

DATE

TIME

WEATHER CONDITIONS

		☀	⛅	☁	🌧	❄
🌡	___					
🚩	___	☐	☐	☐	☐	☐

INSPECTION

HIVE NUMBER	①	②	③	④	⑤	⑥

PRODUCTIVITY & REPRODUCTION

AMOUNT OF HONEY						
GENERAL POPULATION						
AMOUNT OF BROOD						
AMOUNT OF SPACE						

BEHAVIOUR & ACTIVITIES

USUAL ENTERING AND EXITING ACTIVITY?						
CALM BEHAVIOUR WHEN OPENING HIVE?						
BEES BRINGING POLLEN INTO HIVE?						
SIGNS OF ROBBERY AMONG THE BEES?						

HEALTH STATUS

BEES SEEM WEAK OR LAZY?						
HIGH AMOUNT OF DEAD BEES?						
QUEEN BEE IS PRESENT / IDENTIFIABLE?						
INFESTATION BY ANTS / ANTS PRESENT?						
INFESTATION BY WAX MOTH / WAX MOTH PRESENT?						
NEGATIVE ODOUR NOTICEABLE?						

COLONY NAME		WEATHER CONDITIONS					
DATE		🌡 —	☀	⛅	☁	🌧	❄
TIME		🚩 —	☐	☐	☐	☐	☐

INSPECTION

HIVE NUMBER	①	②	③	④	⑤	⑥

PRODUCTIVITY & REPRODUCTION

AMOUNT OF HONEY						
GENERAL POPULATION						
AMOUNT OF BROOD						
AMOUNT OF SPACE						

BEHAVIOUR & ACTIVITIES

USUAL ENTERING AND EXITING ACTIVITY?						
CALM BEHAVIOUR WHEN OPENING HIVE?						
BEES BRINGING POLLEN INTO HIVE?						
SIGNS OF ROBBERY AMONG THE BEES?						

HEALTH STATUS

BEES SEEM WEAK OR LAZY?						
HIGH AMOUNT OF DEAD BEES?						
QUEEN BEE IS PRESENT / IDENTIFIABLE?						
INFESTATION BY ANTS / ANTS PRESENT?						
INFESTATION BY WAX MOTH / WAX MOTH PRESENT?						
NEGATIVE ODOUR NOTICEABLE?						

	COLONY NAME
	DATE
	TIME

WEATHER CONDITIONS

	☼	☁	☁	☂	❄
🌡 ____					
🏳 ____	☐	☐	☐	☐	☐

INSPECTION

HIVE NUMBER	①	②	③	④	⑤	⑥

PRODUCTIVITY & REPRODUCTION

AMOUNT OF HONEY						
GENERAL POPULATION						
AMOUNT OF BROOD						
AMOUNT OF SPACE						

BEHAVIOUR & ACTIVITIES

USUAL ENTERING AND EXITING ACTIVITY?						
CALM BEHAVIOUR WHEN OPENING HIVE?						
BEES BRINGING POLLEN INTO HIVE?						
SIGNS OF ROBBERY AMONG THE BEES?						

HEALTH STATUS

BEES SEEM WEAK OR LAZY?						
HIGH AMOUNT OF DEAD BEES?						
QUEEN BEE IS PRESENT / IDENTIFIABLE?						
INFESTATION BY ANTS / ANTS PRESENT?						
INFESTATION BY WAX MOTH / WAX MOTH PRESENT?						
NEGATIVE ODOUR NOTICEABLE?						

| COLONY NAME | WEATHER CONDITIONS |

COLONY NAME
DATE
TIME

WEATHER CONDITIONS

🌡 —— ☀ ⛅ 🌧 ⛈ ❄

🎐 —— ☐ ☐ ☐ ☐ ☐

INSPECTION

HIVE NUMBER	①	②	③	④	⑤	⑥

PRODUCTIVITY & REPRODUCTION

AMOUNT OF HONEY						
GENERAL POPULATION						
AMOUNT OF BROOD						
AMOUNT OF SPACE						

BEHAVIOUR & ACTIVITIES

USUAL ENTERING AND EXITING ACTIVITY?						
CALM BEHAVIOUR WHEN OPENING HIVE?						
BEES BRINGING POLLEN INTO HIVE?						
SIGNS OF ROBBERY AMONG THE BEES?						

HEALTH STATUS

BEES SEEM WEAK OR LAZY?						
HIGH AMOUNT OF DEAD BEES?						
QUEEN BEE IS PRESENT / IDENTIFIABLE?						
INFESTATION BY ANTS / ANTS PRESENT?						
INFESTATION BY WAX MOTH / WAX MOTH PRESENT?						
NEGATIVE ODOUR NOTICEABLE?						

COLONY NAME		WEATHER CONDITIONS					
DATE		☀	⛅	☁	🌧	❄	
TIME		☐	☐	☐	☐	☐	

INSPECTION

HIVE NUMBER	①	②	③	④	⑤	⑥

PRODUCTIVITY & REPRODUCTION

AMOUNT OF HONEY						
GENERAL POPULATION						
AMOUNT OF BROOD						
AMOUNT OF SPACE						

BEHAVIOUR & ACTIVITIES

USUAL ENTERING AND EXITING ACTIVITY?					
CALM BEHAVIOUR WHEN OPENING HIVE?					
BEES BRINGING POLLEN INTO HIVE?					
SIGNS OF ROBBERY AMONG THE BEES?					

HEALTH STATUS

BEES SEEM WEAK OR LAZY?					
HIGH AMOUNT OF DEAD BEES?					
QUEEN BEE IS PRESENT / IDENTIFIABLE?					
INFESTATION BY ANTS / ANTS PRESENT?					
INFESTATION BY WAX MOTH / WAX MOTH PRESENT?					
NEGATIVE ODOUR NOTICEABLE?					

COLONY NAME		WEATHER CONDITIONS	
DATE			
TIME			

Weather: 🌡 ___ ☀ ⛅ ☁ 🌧 ❄ / 🌬 ___ ☐ ☐ ☐ ☐ ☐

INSPECTION

HIVE NUMBER	①	②	③	④	⑤	⑥

PRODUCTIVITY & REPRODUCTION

AMOUNT OF HONEY						
GENERAL POPULATION						
AMOUNT OF BROOD						
AMOUNT OF SPACE						

BEHAVIOUR & ACTIVITIES

USUAL ENTERING AND EXITING ACTIVITY?						
CALM BEHAVIOUR WHEN OPENING HIVE?						
BEES BRINGING POLLEN INTO HIVE?						
SIGNS OF ROBBERY AMONG THE BEES?						

HEALTH STATUS

BEES SEEM WEAK OR LAZY?						
HIGH AMOUNT OF DEAD BEES?						
QUEEN BEE IS PRESENT / IDENTIFIABLE?						
INFESTATION BY ANTS / ANTS PRESENT?						
INFESTATION BY WAX MOTH / WAX MOTH PRESENT?						
NEGATIVE ODOUR NOTICEABLE?						

COLONY NAME	WEATHER CONDITIONS

COLONY NAME

DATE

TIME

WEATHER CONDITIONS

🌡 —— ☀ ⛅ 🌧 ⛈ ❄

🚩 —— ☐ ☐ ☐ ☐ ☐

INSPECTION

HIVE NUMBER	①	②	③	④	⑤	⑥

PRODUCTIVITY & REPRODUCTION

AMOUNT OF HONEY						
GENERAL POPULATION						
AMOUNT OF BROOD						
AMOUNT OF SPACE						

BEHAVIOUR & ACTIVITIES

USUAL ENTERING AND EXITING ACTIVITY?						
CALM BEHAVIOUR WHEN OPENING HIVE?						
BEES BRINGING POLLEN INTO HIVE?						
SIGNS OF ROBBERY AMONG THE BEES?						

HEALTH STATUS

BEES SEEM WEAK OR LAZY?						
HIGH AMOUNT OF DEAD BEES?						
QUEEN BEE IS PRESENT / IDENTIFIABLE?						
INFESTATION BY ANTS / ANTS PRESENT?						
INFESTATION BY WAX MOTH / WAX MOTH PRESENT?						
NEGATIVE ODOUR NOTICEABLE?						

COLONY NAME		WEATHER CONDITIONS					
DATE		🌡 —	☀	⛅	☁	🌧	❄
TIME		🚩 —	☐	☐	☐	☐	☐

INSPECTION

HIVE NUMBER	①	②	③	④	⑤	⑥

PRODUCTIVITY & REPRODUCTION

AMOUNT OF HONEY						
GENERAL POPULATION						
AMOUNT OF BROOD						
AMOUNT OF SPACE						

BEHAVIOUR & ACTIVITIES

USUAL ENTERING AND EXITING ACTIVITY?						
CALM BEHAVIOUR WHEN OPENING HIVE?						
BEES BRINGING POLLEN INTO HIVE?						
SIGNS OF ROBBERY AMONG THE BEES?						

HEALTH STATUS

BEES SEEM WEAK OR LAZY?						
HIGH AMOUNT OF DEAD BEES?						
QUEEN BEE IS PRESENT / IDENTIFIABLE?						
INFESTATION BY ANTS / ANTS PRESENT?						
INFESTATION BY WAX MOTH / WAX MOTH PRESENT?						
NEGATIVE ODOUR NOTICEABLE?						

COLONY NAME	WEATHER CONDITIONS

COLONY NAME	
DATE	
TIME	

WEATHER CONDITIONS

🌡 —— ☀ ⛅ ☁ 🌧 ❄

🚩 —— ☐ ☐ ☐ ☐ ☐

INSPECTION

HIVE NUMBER	①	②	③	④	⑤	⑥

PRODUCTIVITY & REPRODUCTION

AMOUNT OF HONEY						
GENERAL POPULATION						
AMOUNT OF BROOD						
AMOUNT OF SPACE						

BEHAVIOUR & ACTIVITIES

USUAL ENTERING AND EXITING ACTIVITY?						
CALM BEHAVIOUR WHEN OPENING HIVE?						
BEES BRINGING POLLEN INTO HIVE?						
SIGNS OF ROBBERY AMONG THE BEES?						

HEALTH STATUS

BEES SEEM WEAK OR LAZY?						
HIGH AMOUNT OF DEAD BEES?						
QUEEN BEE IS PRESENT / IDENTIFIABLE?						
INFESTATION BY ANTS / ANTS PRESENT?						
INFESTATION BY WAX MOTH / WAX MOTH PRESENT?						
NEGATIVE ODOUR NOTICEABLE?						

COLONY NAME		WEATHER CONDITIONS					
DATE		🌡 ___	☀	⛅	🌧	⛈	❄
TIME		🚩 ___	☐	☐	☐	☐	☐

INSPECTION

HIVE NUMBER	①	②	③	④	⑤	⑥

PRODUCTIVITY & REPRODUCTION

AMOUNT OF HONEY						
GENERAL POPULATION						
AMOUNT OF BROOD						
AMOUNT OF SPACE						

BEHAVIOUR & ACTIVITIES

USUAL ENTERING AND EXITING ACTIVITY?						
CALM BEHAVIOUR WHEN OPENING HIVE?						
BEES BRINGING POLLEN INTO HIVE?						
SIGNS OF ROBBERY AMONG THE BEES?						

HEALTH STATUS

BEES SEEM WEAK OR LAZY?						
HIGH AMOUNT OF DEAD BEES?						
QUEEN BEE IS PRESENT / IDENTIFIABLE?						
INFESTATION BY ANTS / ANTS PRESENT?						
INFESTATION BY WAX MOTH / WAX MOTH PRESENT?						
NEGATIVE ODOUR NOTICEABLE?						

COLONY NAME		WEATHER CONDITIONS
DATE		☀ ⛅ 🌧 ⛈ ❄
TIME		☐ ☐ ☐ ☐ ☐

INSPECTION

HIVE NUMBER	①	②	③	④	⑤	⑥

PRODUCTIVITY & REPRODUCTION

AMOUNT OF HONEY						
GENERAL POPULATION						
AMOUNT OF BROOD						
AMOUNT OF SPACE						

BEHAVIOUR & ACTIVITIES

USUAL ENTERING AND EXITING ACTIVITY?						
CALM BEHAVIOUR WHEN OPENING HIVE?						
BEES BRINGING POLLEN INTO HIVE?						
SIGNS OF ROBBERY AMONG THE BEES?						

HEALTH STATUS

BEES SEEM WEAK OR LAZY?						
HIGH AMOUNT OF DEAD BEES?						
QUEEN BEE IS PRESENT / IDENTIFIABLE?						
INFESTATION BY ANTS / ANTS PRESENT?						
INFESTATION BY WAX MOTH / WAX MOTH PRESENT?						
NEGATIVE ODOUR NOTICEABLE?						

COLONY NAME		WEATHER CONDITIONS

COLONY NAME

DATE

TIME

WEATHER CONDITIONS

🌡 —— ☀ ⛅ ☁ 🌧 ❄

💨 —— ☐ ☐ ☐ ☐ ☐

INSPECTION

HIVE NUMBER	1	2	3	4	5	6

PRODUCTIVITY & REPRODUCTION

AMOUNT OF HONEY						
GENERAL POPULATION						
AMOUNT OF BROOD						
AMOUNT OF SPACE						

BEHAVIOUR & ACTIVITIES

USUAL ENTERING AND EXITING ACTIVITY?						
CALM BEHAVIOUR WHEN OPENING HIVE?						
BEES BRINGING POLLEN INTO HIVE?						
SIGNS OF ROBBERY AMONG THE BEES?						

HEALTH STATUS

BEES SEEM WEAK OR LAZY?						
HIGH AMOUNT OF DEAD BEES?						
QUEEN BEE IS PRESENT / IDENTIFIABLE?						
INFESTATION BY ANTS / ANTS PRESENT?						
INFESTATION BY WAX MOTH / WAX MOTH PRESENT?						
NEGATIVE ODOUR NOTICEABLE?						

	COLONY NAME	
	DATE	
	TIME	

WEATHER CONDITIONS

🌡️ _____ ☀️ ⛅ 🌧️ ⛈️ ❄️

🚩 _____ ☐ ☐ ☐ ☐ ☐

INSPECTION

HIVE NUMBER	①	②	③	④	⑤	⑥

PRODUCTIVITY & REPRODUCTION

AMOUNT OF HONEY						
GENERAL POPULATION						
AMOUNT OF BROOD						
AMOUNT OF SPACE						

BEHAVIOUR & ACTIVITIES

USUAL ENTERING AND EXITING ACTIVITY?						
CALM BEHAVIOUR WHEN OPENING HIVE?						
BEES BRINGING POLLEN INTO HIVE?						
SIGNS OF ROBBERY AMONG THE BEES?						

HEALTH STATUS

BEES SEEM WEAK OR LAZY?						
HIGH AMOUNT OF DEAD BEES?						
QUEEN BEE IS PRESENT / IDENTIFIABLE?						
INFESTATION BY ANTS / ANTS PRESENT?						
INFESTATION BY WAX MOTH / WAX MOTH PRESENT?						
NEGATIVE ODOUR NOTICEABLE?						

COLONY NAME		WEATHER CONDITIONS					

🌡 ——   ☀ ⛅ 🌧 ⛈ ❄

🚩 ——   ☐ ☐ ☐ ☐ ☐

DATE

TIME

INSPECTION

HIVE NUMBER	①	②	③	④	⑤	⑥

PRODUCTIVITY & REPRODUCTION

	①	②	③	④	⑤	⑥
AMOUNT OF HONEY						
GENERAL POPULATION						
AMOUNT OF BROOD						
AMOUNT OF SPACE						

BEHAVIOUR & ACTIVITIES

	①	②	③	④	⑤	⑥
USUAL ENTERING AND EXITING ACTIVITY?						
CALM BEHAVIOUR WHEN OPENING HIVE?						
BEES BRINGING POLLEN INTO HIVE?						
SIGNS OF ROBBERY AMONG THE BEES?						

HEALTH STATUS

	①	②	③	④	⑤	⑥
BEES SEEM WEAK OR LAZY?						
HIGH AMOUNT OF DEAD BEES?						
QUEEN BEE IS PRESENT / IDENTIFIABLE?						
INFESTATION BY ANTS / ANTS PRESENT?						
INFESTATION BY WAX MOTH / WAX MOTH PRESENT?						
NEGATIVE ODOUR NOTICEABLE?						

COLONY NAME	WEATHER CONDITIONS

COLONY NAME	
📅 DATE	
🕐 TIME	

WEATHER CONDITIONS

🌡 —— ☀️ ⛅ 🌥 🌧 ❄️

🚩 —— ☐ ☐ ☐ ☐ ☐

INSPECTION

HIVE NUMBER	1	2	3	4	5	6

PRODUCTIVITY & REPRODUCTION

AMOUNT OF HONEY						
GENERAL POPULATION						
AMOUNT OF BROOD						
AMOUNT OF SPACE						

BEHAVIOUR & ACTIVITIES

USUAL ENTERING AND EXITING ACTIVITY?						
CALM BEHAVIOUR WHEN OPENING HIVE?						
BEES BRINGING POLLEN INTO HIVE?						
SIGNS OF ROBBERY AMONG THE BEES?						

HEALTH STATUS

BEES SEEM WEAK OR LAZY?						
HIGH AMOUNT OF DEAD BEES?						
QUEEN BEE IS PRESENT / IDENTIFIABLE?						
INFESTATION BY ANTS / ANTS PRESENT?						
INFESTATION BY WAX MOTH / WAX MOTH PRESENT?						
NEGATIVE ODOUR NOTICEABLE?						

COLONY NAME		WEATHER CONDITIONS					
DATE		🌡 ___	☀	⛅	🌧	⛈	❄
TIME		🚩 ___	☐	☐	☐	☐	☐

INSPECTION

HIVE NUMBER	①	②	③	④	⑤	⑥

PRODUCTIVITY & REPRODUCTION

AMOUNT OF HONEY						
GENERAL POPULATION						
AMOUNT OF BROOD						
AMOUNT OF SPACE						

BEHAVIOUR & ACTIVITIES

USUAL ENTERING AND EXITING ACTIVITY?						
CALM BEHAVIOUR WHEN OPENING HIVE?						
BEES BRINGING POLLEN INTO HIVE?						
SIGNS OF ROBBERY AMONG THE BEES?						

HEALTH STATUS

BEES SEEM WEAK OR LAZY?						
HIGH AMOUNT OF DEAD BEES?						
QUEEN BEE IS PRESENT / IDENTIFIABLE?						
INFESTATION BY ANTS / ANTS PRESENT?						
INFESTATION BY WAX MOTH / WAX MOTH PRESENT?						
NEGATIVE ODOUR NOTICEABLE?						

COLONY NAME		WEATHER CONDITIONS					
DATE		🌡 ___	☀	⛅	☁	🌧	❄
TIME		🚩 ___	☐	☐	☐	☐	☐

INSPECTION

HIVE NUMBER	1	2	3	4	5	6

PRODUCTIVITY & REPRODUCTION

AMOUNT OF HONEY						
GENERAL POPULATION						
AMOUNT OF BROOD						
AMOUNT OF SPACE						

BEHAVIOUR & ACTIVITIES

USUAL ENTERING AND EXITING ACTIVITY?						
CALM BEHAVIOUR WHEN OPENING HIVE?						
BEES BRINGING POLLEN INTO HIVE?						
SIGNS OF ROBBERY AMONG THE BEES?						

HEALTH STATUS

BEES SEEM WEAK OR LAZY?						
HIGH AMOUNT OF DEAD BEES?						
QUEEN BEE IS PRESENT / IDENTIFIABLE?						
INFESTATION BY ANTS / ANTS PRESENT?						
INFESTATION BY WAX MOTH / WAX MOTH PRESENT?						
NEGATIVE ODOUR NOTICEABLE?						

COLONY NAME	WEATHER CONDITIONS

COLONY NAME

DATE

TIME

WEATHER CONDITIONS

🌡 ___ ☀ ⛅ ☁ 🌧 ❄

🚩 ___ ☐ ☐ ☐ ☐ ☐

INSPECTION

HIVE NUMBER	①	②	③	④	⑤	⑥

PRODUCTIVITY & REPRODUCTION

AMOUNT OF HONEY						
GENERAL POPULATION						
AMOUNT OF BROOD						
AMOUNT OF SPACE						

BEHAVIOUR & ACTIVITIES

USUAL ENTERING AND EXITING ACTIVITY?						
CALM BEHAVIOUR WHEN OPENING HIVE?						
BEES BRINGING POLLEN INTO HIVE?						
SIGNS OF ROBBERY AMONG THE BEES?						

HEALTH STATUS

BEES SEEM WEAK OR LAZY?						
HIGH AMOUNT OF DEAD BEES?						
QUEEN BEE IS PRESENT / IDENTIFIABLE?						
INFESTATION BY ANTS / ANTS PRESENT?						
INFESTATION BY WAX MOTH / WAX MOTH PRESENT?						
NEGATIVE ODOUR NOTICEABLE?						

COLONY NAME	WEATHER CONDITIONS

COLONY NAME					
DATE					
TIME					

WEATHER CONDITIONS

🌡 ____ ☀ ⛅ 🌧 ⛈ ❄

🏳 ____ ☐ ☐ ☐ ☐ ☐

INSPECTION

HIVE NUMBER	①	②	③	④	⑤	⑥

PRODUCTIVITY & REPRODUCTION

AMOUNT OF HONEY						
GENERAL POPULATION						
AMOUNT OF BROOD						
AMOUNT OF SPACE						

BEHAVIOUR & ACTIVITIES

USUAL ENTERING AND EXITING ACTIVITY?						
CALM BEHAVIOUR WHEN OPENING HIVE?						
BEES BRINGING POLLEN INTO HIVE?						
SIGNS OF ROBBERY AMONG THE BEES?						

HEALTH STATUS

BEES SEEM WEAK OR LAZY?						
HIGH AMOUNT OF DEAD BEES?						
QUEEN BEE IS PRESENT / IDENTIFIABLE?						
INFESTATION BY ANTS / ANTS PRESENT?						
INFESTATION BY WAX MOTH / WAX MOTH PRESENT?						
NEGATIVE ODOUR NOTICEABLE?						

COLONY NAME		WEATHER CONDITIONS					
DATE		🌡 —	☀	⛅	☁	🌧	❄
TIME		🎏 —	☐	☐	☐	☐	☐

INSPECTION

HIVE NUMBER	①	②	③	④	⑤	⑥

PRODUCTIVITY & REPRODUCTION

	1	2	3	4	5	6
AMOUNT OF HONEY						
GENERAL POPULATION						
AMOUNT OF BROOD						
AMOUNT OF SPACE						

BEHAVIOUR & ACTIVITIES

	1	2	3	4	5	6
USUAL ENTERING AND EXITING ACTIVITY?						
CALM BEHAVIOUR WHEN OPENING HIVE?						
BEES BRINGING POLLEN INTO HIVE?						
SIGNS OF ROBBERY AMONG THE BEES?						

HEALTH STATUS

	1	2	3	4	5	6
BEES SEEM WEAK OR LAZY?						
HIGH AMOUNT OF DEAD BEES?						
QUEEN BEE IS PRESENT / IDENTIFIABLE?						
INFESTATION BY ANTS / ANTS PRESENT?						
INFESTATION BY WAX MOTH / WAX MOTH PRESENT?						
NEGATIVE ODOUR NOTICEABLE?						

COLONY NAME		WEATHER CONDITIONS					
DATE		🌡 ____	☀	⛅	🌧	⛈	❄
TIME		🚩 ____	☐	☐	☐	☐	☐

INSPECTION

HIVE NUMBER	①	②	③	④	⑤	⑥

PRODUCTIVITY & REPRODUCTION

AMOUNT OF HONEY						
GENERAL POPULATION						
AMOUNT OF BROOD						
AMOUNT OF SPACE						

BEHAVIOUR & ACTIVITIES

USUAL ENTERING AND EXITING ACTIVITY?						
CALM BEHAVIOUR WHEN OPENING HIVE?						
BEES BRINGING POLLEN INTO HIVE?						
SIGNS OF ROBBERY AMONG THE BEES?						

HEALTH STATUS

BEES SEEM WEAK OR LAZY?						
HIGH AMOUNT OF DEAD BEES?						
QUEEN BEE IS PRESENT / IDENTIFIABLE?						
INFESTATION BY ANTS / ANTS PRESENT?						
INFESTATION BY WAX MOTH / WAX MOTH PRESENT?						
NEGATIVE ODOUR NOTICEABLE?						

COLONY NAME		WEATHER CONDITIONS					
DATE		🌡 ___	☀	⛅	🌧	⛈	❄
TIME		🚩 ___	☐	☐	☐	☐	☐

INSPECTION

HIVE NUMBER	1	2	3	4	5	6

PRODUCTIVITY & REPRODUCTION

AMOUNT OF HONEY						
GENERAL POPULATION						
AMOUNT OF BROOD						
AMOUNT OF SPACE						

BEHAVIOUR & ACTIVITIES

USUAL ENTERING AND EXITING ACTIVITY?						
CALM BEHAVIOUR WHEN OPENING HIVE?						
BEES BRINGING POLLEN INTO HIVE?						
SIGNS OF ROBBERY AMONG THE BEES?						

HEALTH STATUS

BEES SEEM WEAK OR LAZY?						
HIGH AMOUNT OF DEAD BEES?						
QUEEN BEE IS PRESENT / IDENTIFIABLE?						
INFESTATION BY ANTS / ANTS PRESENT?						
INFESTATION BY WAX MOTH / WAX MOTH PRESENT?						
NEGATIVE ODOUR NOTICEABLE?						

COLONY NAME	
DATE	
TIME	

WEATHER CONDITIONS

🌡 ___ ☀ ⛅ ☁ 🌧 ❄

🚩 ___ ☐ ☐ ☐ ☐ ☐

INSPECTION

HIVE NUMBER	(1)	(2)	(3)	(4)	(5)	(6)

PRODUCTIVITY & REPRODUCTION

AMOUNT OF HONEY						
GENERAL POPULATION						
AMOUNT OF BROOD						
AMOUNT OF SPACE						

BEHAVIOUR & ACTIVITIES

USUAL ENTERING AND EXITING ACTIVITY?						
CALM BEHAVIOUR WHEN OPENING HIVE?						
BEES BRINGING POLLEN INTO HIVE?						
SIGNS OF ROBBERY AMONG THE BEES?						

HEALTH STATUS

BEES SEEM WEAK OR LAZY?						
HIGH AMOUNT OF DEAD BEES?						
QUEEN BEE IS PRESENT / IDENTIFIABLE?						
INFESTATION BY ANTS / ANTS PRESENT?						
INFESTATION BY WAX MOTH / WAX MOTH PRESENT?						
NEGATIVE ODOUR NOTICEABLE?						

COLONY NAME		WEATHER CONDITIONS
DATE		🌡 ____ ☀ ⛅ ☁ 🌧 ❄
TIME		🚩 ____ ☐ ☐ ☐ ☐ ☐

INSPECTION

HIVE NUMBER	①	②	③	④	⑤	⑥

PRODUCTIVITY & REPRODUCTION

AMOUNT OF HONEY						
GENERAL POPULATION						
AMOUNT OF BROOD						
AMOUNT OF SPACE						

BEHAVIOUR & ACTIVITIES

USUAL ENTERING AND EXITING ACTIVITY?						
CALM BEHAVIOUR WHEN OPENING HIVE?						
BEES BRINGING POLLEN INTO HIVE?						
SIGNS OF ROBBERY AMONG THE BEES?						

HEALTH STATUS

BEES SEEM WEAK OR LAZY?						
HIGH AMOUNT OF DEAD BEES?						
QUEEN BEE IS PRESENT / IDENTIFIABLE?						
INFESTATION BY ANTS / ANTS PRESENT?						
INFESTATION BY WAX MOTH / WAX MOTH PRESENT?						
NEGATIVE ODOUR NOTICEABLE?						

COLONY NAME					WEATHER CONDITIONS				

COLONY NAME

DATE

TIME

WEATHER CONDITIONS

🌡 ___ ☀ ⛅ ☁ 🌧 ❄

🚩 ___ ☐ ☐ ☐ ☐ ☐

INSPECTION

HIVE NUMBER	1	2	3	4	5	6

PRODUCTIVITY & REPRODUCTION

AMOUNT OF HONEY						
GENERAL POPULATION						
AMOUNT OF BROOD						
AMOUNT OF SPACE						

BEHAVIOUR & ACTIVITIES

USUAL ENTERING AND EXITING ACTIVITY?						
CALM BEHAVIOUR WHEN OPENING HIVE?						
BEES BRINGING POLLEN INTO HIVE?						
SIGNS OF ROBBERY AMONG THE BEES?						

HEALTH STATUS

BEES SEEM WEAK OR LAZY?						
HIGH AMOUNT OF DEAD BEES?						
QUEEN BEE IS PRESENT / IDENTIFIABLE?						
INFESTATION BY ANTS / ANTS PRESENT?						
INFESTATION BY WAX MOTH / WAX MOTH PRESENT?						
NEGATIVE ODOUR NOTICEABLE?						

COLONY NAME		WEATHER CONDITIONS
DATE		
TIME		

Weather icons: 🌡 — ☀ ⛅ ☁ 🌧 ❄
Wind: — ☐ ☐ ☐ ☐ ☐

INSPECTION

HIVE NUMBER	①	②	③	④	⑤	⑥

PRODUCTIVITY & REPRODUCTION

AMOUNT OF HONEY						
GENERAL POPULATION						
AMOUNT OF BROOD						
AMOUNT OF SPACE						

BEHAVIOUR & ACTIVITIES

USUAL ENTERING AND EXITING ACTIVITY?						
CALM BEHAVIOUR WHEN OPENING HIVE?						
BEES BRINGING POLLEN INTO HIVE?						
SIGNS OF ROBBERY AMONG THE BEES?						

HEALTH STATUS

BEES SEEM WEAK OR LAZY?						
HIGH AMOUNT OF DEAD BEES?						
QUEEN BEE IS PRESENT / IDENTIFIABLE?						
INFESTATION BY ANTS / ANTS PRESENT?						
INFESTATION BY WAX MOTH / WAX MOTH PRESENT?						
NEGATIVE ODOUR NOTICEABLE?						

<table>
<tr><td colspan="2">

🏠 **COLONY NAME**

📅 **DATE**

🕐 **TIME**
</td><td colspan="2">

WEATHER CONDITIONS

🌡️ ___ ☀️ ⛅ 🌧️ ⛈️ ❄️

🚩 ___ ☐ ☐ ☐ ☐ ☐
</td></tr>
</table>

INSPECTION

🐝 HIVE NUMBER	①	②	③	④	⑤	⑥

PRODUCTIVITY & REPRODUCTION

🍯 AMOUNT OF HONEY						
🐝 GENERAL POPULATION						
🐝 AMOUNT OF BROOD						
🍯 AMOUNT OF SPACE						

BEHAVIOUR & ACTIVITIES

🐝 USUAL ENTERING AND EXITING ACTIVITY?						
📦 CALM BEHAVIOUR WHEN OPENING HIVE?						
🐝 BEES BRINGING POLLEN INTO HIVE?						
🍯 SIGNS OF ROBBERY AMONG THE BEES?						

HEALTH STATUS

🐝 BEES SEEM WEAK OR LAZY?						
☠️ HIGH AMOUNT OF DEAD BEES?						
🐝 QUEEN BEE IS PRESENT / IDENTIFIABLE?						
🐜 INFESTATION BY ANTS / ANTS PRESENT?						
🦋 INFESTATION BY WAX MOTH / WAX MOTH PRESENT?						
👃 NEGATIVE ODOUR NOTICEABLE?						

COLONY NAME		WEATHER CONDITIONS					
DATE		🌡 ———	☀	⛅	🌧	⛈	❄
TIME		🚩 ———	☐	☐	☐	☐	☐

INSPECTION

HIVE NUMBER	①	②	③	④	⑤	⑥

PRODUCTIVITY & REPRODUCTION

AMOUNT OF HONEY						
GENERAL POPULATION						
AMOUNT OF BROOD						
AMOUNT OF SPACE						

BEHAVIOUR & ACTIVITIES

USUAL ENTERING AND EXITING ACTIVITY?						
CALM BEHAVIOUR WHEN OPENING HIVE?						
BEES BRINGING POLLEN INTO HIVE?						
SIGNS OF ROBBERY AMONG THE BEES?						

HEALTH STATUS

BEES SEEM WEAK OR LAZY?						
HIGH AMOUNT OF DEAD BEES?						
QUEEN BEE IS PRESENT / IDENTIFIABLE?						
INFESTATION BY ANTS / ANTS PRESENT?						
INFESTATION BY WAX MOTH / WAX MOTH PRESENT?						
NEGATIVE ODOUR NOTICEABLE?						

COLONY NAME		WEATHER CONDITIONS					
DATE		🌡 ___	☀	⛅	☁	🌧	❄
TIME		🚩 ___	☐	☐	☐	☐	☐

INSPECTION

HIVE NUMBER	1	2	3	4	5	6

PRODUCTIVITY & REPRODUCTION

AMOUNT OF HONEY						
GENERAL POPULATION						
AMOUNT OF BROOD						
AMOUNT OF SPACE						

BEHAVIOUR & ACTIVITIES

USUAL ENTERING AND EXITING ACTIVITY?					
CALM BEHAVIOUR WHEN OPENING HIVE?					
BEES BRINGING POLLEN INTO HIVE?					
SIGNS OF ROBBERY AMONG THE BEES?					

HEALTH STATUS

BEES SEEM WEAK OR LAZY?					
HIGH AMOUNT OF DEAD BEES?					
QUEEN BEE IS PRESENT / IDENTIFIABLE?					
INFESTATION BY ANTS / ANTS PRESENT?					
INFESTATION BY WAX MOTH / WAX MOTH PRESENT?					
NEGATIVE ODOUR NOTICEABLE?					

	COLONY NAME	
	DATE	
	TIME	

WEATHER CONDITIONS

🌡 ___   ☀ ⛅ 🌧 ⛈ ❄

🚩 ___   ☐ ☐ ☐ ☐ ☐

INSPECTION

	HIVE NUMBER	①	②	③	④	⑤	⑥

PRODUCTIVITY & REPRODUCTION

	AMOUNT OF HONEY						
	GENERAL POPULATION						
	AMOUNT OF BROOD						
	AMOUNT OF SPACE						

BEHAVIOUR & ACTIVITIES

	USUAL ENTERING AND EXITING ACTIVITY?						
	CALM BEHAVIOUR WHEN OPENING HIVE?						
	BEES BRINGING POLLEN INTO HIVE?						
	SIGNS OF ROBBERY AMONG THE BEES?						

HEALTH STATUS

	BEES SEEM WEAK OR LAZY?						
	HIGH AMOUNT OF DEAD BEES?						
	QUEEN BEE IS PRESENT / IDENTIFIABLE?						
	INFESTATION BY ANTS / ANTS PRESENT?						
	INFESTATION BY WAX MOTH / WAX MOTH PRESENT?						
	NEGATIVE ODOUR NOTICEABLE?						

COLONY NAME		WEATHER CONDITIONS					
DATE		🌡 ___	☀	⛅	🌧	⛈	❄
TIME		🚩 ___	☐	☐	☐	☐	☐

INSPECTION

HIVE NUMBER	①	②	③	④	⑤	⑥

PRODUCTIVITY & REPRODUCTION

AMOUNT OF HONEY						
GENERAL POPULATION						
AMOUNT OF BROOD						
AMOUNT OF SPACE						

BEHAVIOUR & ACTIVITIES

USUAL ENTERING AND EXITING ACTIVITY?						
CALM BEHAVIOUR WHEN OPENING HIVE?						
BEES BRINGING POLLEN INTO HIVE?						
SIGNS OF ROBBERY AMONG THE BEES?						

HEALTH STATUS

BEES SEEM WEAK OR LAZY?						
HIGH AMOUNT OF DEAD BEES?						
QUEEN BEE IS PRESENT / IDENTIFIABLE?						
INFESTATION BY ANTS / ANTS PRESENT?						
INFESTATION BY WAX MOTH / WAX MOTH PRESENT?						
NEGATIVE ODOUR NOTICEABLE?						

COLONY NAME		WEATHER CONDITIONS					
DATE		🌡 ___	☀	⛅	☁	🌧	❄
TIME		🎐 ___	☐	☐	☐	☐	☐

INSPECTION

	1	2	3	4	5	6
HIVE NUMBER	①	②	③	④	⑤	⑥

PRODUCTIVITY & REPRODUCTION

AMOUNT OF HONEY						
GENERAL POPULATION						
AMOUNT OF BROOD						
AMOUNT OF SPACE						

BEHAVIOUR & ACTIVITIES

USUAL ENTERING AND EXITING ACTIVITY?						
CALM BEHAVIOUR WHEN OPENING HIVE?						
BEES BRINGING POLLEN INTO HIVE?						
SIGNS OF ROBBERY AMONG THE BEES?						

HEALTH STATUS

BEES SEEM WEAK OR LAZY?						
HIGH AMOUNT OF DEAD BEES?						
QUEEN BEE IS PRESENT / IDENTIFIABLE?						
INFESTATION BY ANTS / ANTS PRESENT?						
INFESTATION BY WAX MOTH / WAX MOTH PRESENT?						
NEGATIVE ODOUR NOTICEABLE?						

	COLONY NAME
	DATE
	TIME

WEATHER CONDITIONS

🌡 ___ ☀ ⛅ 🌧 ⛈ ❄

🚩 ___ ☐ ☐ ☐ ☐ ☐

INSPECTION

	HIVE NUMBER	①	②	③	④	⑤	⑥

PRODUCTIVITY & REPRODUCTION

	AMOUNT OF HONEY						
	GENERAL POPULATION						
	AMOUNT OF BROOD						
	AMOUNT OF SPACE						

BEHAVIOUR & ACTIVITIES

	USUAL ENTERING AND EXITING ACTIVITY?						
	CALM BEHAVIOUR WHEN OPENING HIVE?						
	BEES BRINGING POLLEN INTO HIVE?						
	SIGNS OF ROBBERY AMONG THE BEES?						

HEALTH STATUS

	BEES SEEM WEAK OR LAZY?						
	HIGH AMOUNT OF DEAD BEES?						
	QUEEN BEE IS PRESENT / IDENTIFIABLE?						
	INFESTATION BY ANTS / ANTS PRESENT?						
	INFESTATION BY WAX MOTH / WAX MOTH PRESENT?						
	NEGATIVE ODOUR NOTICEABLE?						

COLONY NAME		WEATHER CONDITIONS					
DATE		🌡 ___	☀	⛅	☁	🌧	❄
TIME		🚩 ___	☐	☐	☐	☐	☐

INSPECTION

HIVE NUMBER	①	②	③	④	⑤	⑥

PRODUCTIVITY & REPRODUCTION

AMOUNT OF HONEY						
GENERAL POPULATION						
AMOUNT OF BROOD						
AMOUNT OF SPACE						

BEHAVIOUR & ACTIVITIES

USUAL ENTERING AND EXITING ACTIVITY?						
CALM BEHAVIOUR WHEN OPENING HIVE?						
BEES BRINGING POLLEN INTO HIVE?						
SIGNS OF ROBBERY AMONG THE BEES?						

HEALTH STATUS

BEES SEEM WEAK OR LAZY?						
HIGH AMOUNT OF DEAD BEES?						
QUEEN BEE IS PRESENT / IDENTIFIABLE?						
INFESTATION BY ANTS / ANTS PRESENT?						
INFESTATION BY WAX MOTH / WAX MOTH PRESENT?						
NEGATIVE ODOUR NOTICEABLE?						

COLONY NAME	
DATE	
TIME	

WEATHER CONDITIONS

🌡 ____   ☀ ⛅ ☁ 🌧 ❄

🎏 ____   ☐ ☐ ☐ ☐ ☐

INSPECTION

🐝 HIVE NUMBER	①	②	③	④	⑤	⑥

PRODUCTIVITY & REPRODUCTION

AMOUNT OF HONEY						
GENERAL POPULATION						
AMOUNT OF BROOD						
AMOUNT OF SPACE						

BEHAVIOUR & ACTIVITIES

USUAL ENTERING AND EXITING ACTIVITY?						
CALM BEHAVIOUR WHEN OPENING HIVE?						
BEES BRINGING POLLEN INTO HIVE?						
SIGNS OF ROBBERY AMONG THE BEES?						

HEALTH STATUS

BEES SEEM WEAK OR LAZY?						
HIGH AMOUNT OF DEAD BEES?						
QUEEN BEE IS PRESENT / IDENTIFIABLE?						
INFESTATION BY ANTS / ANTS PRESENT?						
INFESTATION BY WAX MOTH / WAX MOTH PRESENT?						
NEGATIVE ODOUR NOTICEABLE?						

COLONY NAME		WEATHER CONDITIONS					
DATE		🌡 ___	☀	⛅	☁	🌧	❄
TIME		🚩 ___	☐	☐	☐	☐	☐

INSPECTION

HIVE NUMBER	①	②	③	④	⑤	⑥

PRODUCTIVITY & REPRODUCTION

AMOUNT OF HONEY						
GENERAL POPULATION						
AMOUNT OF BROOD						
AMOUNT OF SPACE						

BEHAVIOUR & ACTIVITIES

USUAL ENTERING AND EXITING ACTIVITY?						
CALM BEHAVIOUR WHEN OPENING HIVE?						
BEES BRINGING POLLEN INTO HIVE?						
SIGNS OF ROBBERY AMONG THE BEES?						

HEALTH STATUS

BEES SEEM WEAK OR LAZY?						
HIGH AMOUNT OF DEAD BEES?						
QUEEN BEE IS PRESENT / IDENTIFIABLE?						
INFESTATION BY ANTS / ANTS PRESENT?						
INFESTATION BY WAX MOTH / WAX MOTH PRESENT?						
NEGATIVE ODOUR NOTICEABLE?						

COLONY NAME	WEATHER CONDITIONS
DATE	☼ ⛅ ☁ ⛈ ❄
TIME	☐ ☐ ☐ ☐ ☐

INSPECTION

HIVE NUMBER	①	②	③	④	⑤	⑥

PRODUCTIVITY & REPRODUCTION

AMOUNT OF HONEY						
GENERAL POPULATION						
AMOUNT OF BROOD						
AMOUNT OF SPACE						

BEHAVIOUR & ACTIVITIES

USUAL ENTERING AND EXITING ACTIVITY?						
CALM BEHAVIOUR WHEN OPENING HIVE?						
BEES BRINGING POLLEN INTO HIVE?						
SIGNS OF ROBBERY AMONG THE BEES?						

HEALTH STATUS

BEES SEEM WEAK OR LAZY?						
HIGH AMOUNT OF DEAD BEES?						
QUEEN BEE IS PRESENT / IDENTIFIABLE?						
INFESTATION BY ANTS / ANTS PRESENT?						
INFESTATION BY WAX MOTH / WAX MOTH PRESENT?						
NEGATIVE ODOUR NOTICEABLE?						

COLONY NAME		WEATHER CONDITIONS					
DATE		🌡 — ☀ ⛅ 🌧 ⛈ ❄					
TIME		📶 — ☐ ☐ ☐ ☐ ☐					

INSPECTION

HIVE NUMBER	①	②	③	④	⑤	⑥

PRODUCTIVITY & REPRODUCTION

AMOUNT OF HONEY						
GENERAL POPULATION						
AMOUNT OF BROOD						
AMOUNT OF SPACE						

BEHAVIOUR & ACTIVITIES

USUAL ENTERING AND EXITING ACTIVITY?						
CALM BEHAVIOUR WHEN OPENING HIVE?						
BEES BRINGING POLLEN INTO HIVE?						
SIGNS OF ROBBERY AMONG THE BEES?						

HEALTH STATUS

BEES SEEM WEAK OR LAZY?						
HIGH AMOUNT OF DEAD BEES?						
QUEEN BEE IS PRESENT / IDENTIFIABLE?						
INFESTATION BY ANTS / ANTS PRESENT?						
INFESTATION BY WAX MOTH / WAX MOTH PRESENT?						
NEGATIVE ODOUR NOTICEABLE?						

	COLONY NAME
	DATE
	TIME

WEATHER CONDITIONS

🌡 ____ ☀ ⛅ 🌧 ⛈ ❄

🚩 ____ ☐ ☐ ☐ ☐ ☐

INSPECTION

HIVE NUMBER	①	②	③	④	⑤	⑥

PRODUCTIVITY & REPRODUCTION

AMOUNT OF HONEY						
GENERAL POPULATION						
AMOUNT OF BROOD						
AMOUNT OF SPACE						

BEHAVIOUR & ACTIVITIES

USUAL ENTERING AND EXITING ACTIVITY?						
CALM BEHAVIOUR WHEN OPENING HIVE?						
BEES BRINGING POLLEN INTO HIVE?						
SIGNS OF ROBBERY AMONG THE BEES?						

HEALTH STATUS

BEES SEEM WEAK OR LAZY?						
HIGH AMOUNT OF DEAD BEES?						
QUEEN BEE IS PRESENT / IDENTIFIABLE?						
INFESTATION BY ANTS / ANTS PRESENT?						
INFESTATION BY WAX MOTH / WAX MOTH PRESENT?						
NEGATIVE ODOUR NOTICEABLE?						

COLONY NAME		WEATHER CONDITIONS					
DATE		🌡 ____	☀	⛅	🌧	⛈	❄
TIME		🚩 ____	☐	☐	☐	☐	☐

INSPECTION

HIVE NUMBER	①	②	③	④	⑤	⑥

PRODUCTIVITY & REPRODUCTION

	1	2	3	4	5	6
AMOUNT OF HONEY						
GENERAL POPULATION						
AMOUNT OF BROOD						
AMOUNT OF SPACE						

BEHAVIOUR & ACTIVITIES

	1	2	3	4	5	6
USUAL ENTERING AND EXITING ACTIVITY?						
CALM BEHAVIOUR WHEN OPENING HIVE?						
BEES BRINGING POLLEN INTO HIVE?						
SIGNS OF ROBBERY AMONG THE BEES?						

HEALTH STATUS

	1	2	3	4	5	6
BEES SEEM WEAK OR LAZY?						
HIGH AMOUNT OF DEAD BEES?						
QUEEN BEE IS PRESENT / IDENTIFIABLE?						
INFESTATION BY ANTS / ANTS PRESENT?						
INFESTATION BY WAX MOTH / WAX MOTH PRESENT?						
NEGATIVE ODOUR NOTICEABLE?						

COLONY NAME	WEATHER CONDITIONS

COLONY NAME	
DATE	
TIME	

WEATHER CONDITIONS

🌡 ___ ☀ ⛅ ☁ 🌧 ❄

🚩 ___ ☐ ☐ ☐ ☐ ☐

INSPECTION

HIVE NUMBER	①	②	③	④	⑤	⑥

PRODUCTIVITY & REPRODUCTION

AMOUNT OF HONEY						
GENERAL POPULATION						
AMOUNT OF BROOD						
AMOUNT OF SPACE						

BEHAVIOUR & ACTIVITIES

USUAL ENTERING AND EXITING ACTIVITY?						
CALM BEHAVIOUR WHEN OPENING HIVE?						
BEES BRINGING POLLEN INTO HIVE?						
SIGNS OF ROBBERY AMONG THE BEES?						

HEALTH STATUS

BEES SEEM WEAK OR LAZY?						
HIGH AMOUNT OF DEAD BEES?						
QUEEN BEE IS PRESENT / IDENTIFIABLE?						
INFESTATION BY ANTS / ANTS PRESENT?						
INFESTATION BY WAX MOTH / WAX MOTH PRESENT?						
NEGATIVE ODOUR NOTICEABLE?						

COLONY NAME		WEATHER CONDITIONS					

COLONY NAME	
DATE	
TIME	

WEATHER CONDITIONS

🌡 ____   ☀ ⛅ 🌧 ⛈ ❄

🚩 ____   ☐ ☐ ☐ ☐ ☐

INSPECTION

HIVE NUMBER	1	2	3	4	5	6

PRODUCTIVITY & REPRODUCTION

	1	2	3	4	5	6
AMOUNT OF HONEY						
GENERAL POPULATION						
AMOUNT OF BROOD						
AMOUNT OF SPACE						

BEHAVIOUR & ACTIVITIES

	1	2	3	4	5	6
USUAL ENTERING AND EXITING ACTIVITY?						
CALM BEHAVIOUR WHEN OPENING HIVE?						
BEES BRINGING POLLEN INTO HIVE?						
SIGNS OF ROBBERY AMONG THE BEES?						

HEALTH STATUS

	1	2	3	4	5	6
BEES SEEM WEAK OR LAZY?						
HIGH AMOUNT OF DEAD BEES?						
QUEEN BEE IS PRESENT / IDENTIFIABLE?						
INFESTATION BY ANTS / ANTS PRESENT?						
INFESTATION BY WAX MOTH / WAX MOTH PRESENT?						
NEGATIVE ODOUR NOTICEABLE?						

COLONY NAME		WEATHER CONDITIONS					
DATE		🌡 ___	☀	⛅	🌧	⛈	❄
TIME		🚩 ___	☐	☐	☐	☐	☐

INSPECTION

HIVE NUMBER	1	2	3	4	5	6

PRODUCTIVITY & REPRODUCTION

AMOUNT OF HONEY						
GENERAL POPULATION						
AMOUNT OF BROOD						
AMOUNT OF SPACE						

BEHAVIOUR & ACTIVITIES

USUAL ENTERING AND EXITING ACTIVITY?						
CALM BEHAVIOUR WHEN OPENING HIVE?						
BEES BRINGING POLLEN INTO HIVE?						
SIGNS OF ROBBERY AMONG THE BEES?						

HEALTH STATUS

BEES SEEM WEAK OR LAZY?						
HIGH AMOUNT OF DEAD BEES?						
QUEEN BEE IS PRESENT / IDENTIFIABLE?						
INFESTATION BY ANTS / ANTS PRESENT?						
INFESTATION BY WAX MOTH / WAX MOTH PRESENT?						
NEGATIVE ODOUR NOTICEABLE?						

COLONY NAME	WEATHER CONDITIONS

COLONY NAME

DATE

TIME

WEATHER CONDITIONS

INSPECTION

HIVE NUMBER	①	②	③	④	⑤	⑥

PRODUCTIVITY & REPRODUCTION

AMOUNT OF HONEY						
GENERAL POPULATION						
AMOUNT OF BROOD						
AMOUNT OF SPACE						

BEHAVIOUR & ACTIVITIES

USUAL ENTERING AND EXITING ACTIVITY?						
CALM BEHAVIOUR WHEN OPENING HIVE?						
BEES BRINGING POLLEN INTO HIVE?						
SIGNS OF ROBBERY AMONG THE BEES?						

HEALTH STATUS

BEES SEEM WEAK OR LAZY?						
HIGH AMOUNT OF DEAD BEES?						
QUEEN BEE IS PRESENT / IDENTIFIABLE?						
INFESTATION BY ANTS / ANTS PRESENT?						
INFESTATION BY WAX MOTH / WAX MOTH PRESENT?						
NEGATIVE ODOUR NOTICEABLE?						

COLONY NAME		WEATHER CONDITIONS					
DATE		☀ ⛅ ☁ ⛈ ❄					
TIME		☐ ☐ ☐ ☐ ☐					

INSPECTION

HIVE NUMBER	①	②	③	④	⑤	⑥

PRODUCTIVITY & REPRODUCTION

AMOUNT OF HONEY						
GENERAL POPULATION						
AMOUNT OF BROOD						
AMOUNT OF SPACE						

BEHAVIOUR & ACTIVITIES

USUAL ENTERING AND EXITING ACTIVITY?						
CALM BEHAVIOUR WHEN OPENING HIVE?						
BEES BRINGING POLLEN INTO HIVE?						
SIGNS OF ROBBERY AMONG THE BEES?						

HEALTH STATUS

BEES SEEM WEAK OR LAZY?						
HIGH AMOUNT OF DEAD BEES?						
QUEEN BEE IS PRESENT / IDENTIFIABLE?						
INFESTATION BY ANTS / ANTS PRESENT?						
INFESTATION BY WAX MOTH / WAX MOTH PRESENT?						
NEGATIVE ODOUR NOTICEABLE?						

	COLONY NAME
	DATE
	TIME

WEATHER CONDITIONS

🌡 _____   ☀ ⛅ 🌧 ⛈ ❄

🏳 _____   ☐ ☐ ☐ ☐ ☐

INSPECTION

	HIVE NUMBER	①	②	③	④	⑤	⑥

PRODUCTIVITY & REPRODUCTION

	AMOUNT OF HONEY						
	GENERAL POPULATION						
	AMOUNT OF BROOD						
	AMOUNT OF SPACE						

BEHAVIOUR & ACTIVITIES

	USUAL ENTERING AND EXITING ACTIVITY?						
	CALM BEHAVIOUR WHEN OPENING HIVE?						
	BEES BRINGING POLLEN INTO HIVE?						
	SIGNS OF ROBBERY AMONG THE BEES?						

HEALTH STATUS

	BEES SEEM WEAK OR LAZY?						
	HIGH AMOUNT OF DEAD BEES?						
	QUEEN BEE IS PRESENT / IDENTIFIABLE?						
	INFESTATION BY ANTS / ANTS PRESENT?						
	INFESTATION BY WAX MOTH / WAX MOTH PRESENT?						
	NEGATIVE ODOUR NOTICEABLE?						

	COLONY NAME	
	DATE	
	TIME	

WEATHER CONDITIONS

		☀	⛅	☁	🌧	❄
🌡	——					
🚩	——	☐	☐	☐	☐	☐

INSPECTION

HIVE NUMBER	1	2	3	4	5	6

PRODUCTIVITY & REPRODUCTION

	1	2	3	4	5	6
AMOUNT OF HONEY						
GENERAL POPULATION						
AMOUNT OF BROOD						
AMOUNT OF SPACE						

BEHAVIOUR & ACTIVITIES

USUAL ENTERING AND EXITING ACTIVITY?						
CALM BEHAVIOUR WHEN OPENING HIVE?						
BEES BRINGING POLLEN INTO HIVE?						
SIGNS OF ROBBERY AMONG THE BEES?						

HEALTH STATUS

BEES SEEM WEAK OR LAZY?						
HIGH AMOUNT OF DEAD BEES?						
QUEEN BEE IS PRESENT / IDENTIFIABLE?						
INFESTATION BY ANTS / ANTS PRESENT?						
INFESTATION BY WAX MOTH / WAX MOTH PRESENT?						
NEGATIVE ODOUR NOTICEABLE?						

COLONY NAME	WEATHER CONDITIONS

COLONY NAME		🌡 ___	☀	⛅	🌧	⛈	❄
DATE		🚩 ___	☐	☐	☐	☐	☐
TIME							

INSPECTION

HIVE NUMBER	①	②	③	④	⑤	⑥

PRODUCTIVITY & REPRODUCTION

AMOUNT OF HONEY						
GENERAL POPULATION						
AMOUNT OF BROOD						
AMOUNT OF SPACE						

BEHAVIOUR & ACTIVITIES

USUAL ENTERING AND EXITING ACTIVITY?						
CALM BEHAVIOUR WHEN OPENING HIVE?						
BEES BRINGING POLLEN INTO HIVE?						
SIGNS OF ROBBERY AMONG THE BEES?						

HEALTH STATUS

BEES SEEM WEAK OR LAZY?						
HIGH AMOUNT OF DEAD BEES?						
QUEEN BEE IS PRESENT / IDENTIFIABLE?						
INFESTATION BY ANTS / ANTS PRESENT?						
INFESTATION BY WAX MOTH / WAX MOTH PRESENT?						
NEGATIVE ODOUR NOTICEABLE?						

COLONY NAME		WEATHER CONDITIONS					
DATE		🌡 ___	☀	⛅	🌧	⛈	❄
TIME		🚩 ___	☐	☐	☐	☐	☐

INSPECTION

HIVE NUMBER	①	②	③	④	⑤	⑥

PRODUCTIVITY & REPRODUCTION

AMOUNT OF HONEY						
GENERAL POPULATION						
AMOUNT OF BROOD						
AMOUNT OF SPACE						

BEHAVIOUR & ACTIVITIES

USUAL ENTERING AND EXITING ACTIVITY?						
CALM BEHAVIOUR WHEN OPENING HIVE?						
BEES BRINGING POLLEN INTO HIVE?						
SIGNS OF ROBBERY AMONG THE BEES?						

HEALTH STATUS

BEES SEEM WEAK OR LAZY?						
HIGH AMOUNT OF DEAD BEES?						
QUEEN BEE IS PRESENT / IDENTIFIABLE?						
INFESTATION BY ANTS / ANTS PRESENT?						
INFESTATION BY WAX MOTH / WAX MOTH PRESENT?						
NEGATIVE ODOUR NOTICEABLE?						

	COLONY NAME
	DATE
	TIME

WEATHER CONDITIONS

☀ ⛅ 🌧 ⛈ ❄

☐ ☐ ☐ ☐ ☐

INSPECTION

HIVE NUMBER	①	②	③	④	⑤	⑥

PRODUCTIVITY & REPRODUCTION

AMOUNT OF HONEY						
GENERAL POPULATION						
AMOUNT OF BROOD						
AMOUNT OF SPACE						

BEHAVIOUR & ACTIVITIES

USUAL ENTERING AND EXITING ACTIVITY?						
CALM BEHAVIOUR WHEN OPENING HIVE?						
BEES BRINGING POLLEN INTO HIVE?						
SIGNS OF ROBBERY AMONG THE BEES?						

HEALTH STATUS

BEES SEEM WEAK OR LAZY?						
HIGH AMOUNT OF DEAD BEES?						
QUEEN BEE IS PRESENT / IDENTIFIABLE?						
INFESTATION BY ANTS / ANTS PRESENT?						
INFESTATION BY WAX MOTH / WAX MOTH PRESENT?						
NEGATIVE ODOUR NOTICEABLE?						

COLONY NAME	WEATHER CONDITIONS

COLONY NAME

DATE

TIME

WEATHER CONDITIONS

🌡 ____   ☀ ⛅ ☁ 🌧 ❄

🚩 ____   ☐ ☐ ☐ ☐ ☐

INSPECTION

🐝 HIVE NUMBER	①	②	③	④	⑤	⑥

PRODUCTIVITY & REPRODUCTION

AMOUNT OF HONEY						
GENERAL POPULATION						
AMOUNT OF BROOD						
AMOUNT OF SPACE						

BEHAVIOUR & ACTIVITIES

USUAL ENTERING AND EXITING ACTIVITY?						
CALM BEHAVIOUR WHEN OPENING HIVE?						
BEES BRINGING POLLEN INTO HIVE?						
SIGNS OF ROBBERY AMONG THE BEES?						

HEALTH STATUS

BEES SEEM WEAK OR LAZY?						
HIGH AMOUNT OF DEAD BEES?						
QUEEN BEE IS PRESENT / IDENTIFIABLE?						
INFESTATION BY ANTS / ANTS PRESENT?						
INFESTATION BY WAX MOTH / WAX MOTH PRESENT?						
NEGATIVE ODOUR NOTICEABLE?						

	COLONY NAME
	DATE
	TIME

WEATHER CONDITIONS

🌡	——	☀	⛅	🌧	⛈	❄
🚩	——	☐	☐	☐	☐	☐

INSPECTION

HIVE NUMBER	①	②	③	④	⑤	⑥

PRODUCTIVITY & REPRODUCTION

AMOUNT OF HONEY						
GENERAL POPULATION						
AMOUNT OF BROOD						
AMOUNT OF SPACE						

BEHAVIOUR & ACTIVITIES

USUAL ENTERING AND EXITING ACTIVITY?						
CALM BEHAVIOUR WHEN OPENING HIVE?						
BEES BRINGING POLLEN INTO HIVE?						
SIGNS OF ROBBERY AMONG THE BEES?						

HEALTH STATUS

BEES SEEM WEAK OR LAZY?						
HIGH AMOUNT OF DEAD BEES?						
QUEEN BEE IS PRESENT / IDENTIFIABLE?						
INFESTATION BY ANTS / ANTS PRESENT?						
INFESTATION BY WAX MOTH / WAX MOTH PRESENT?						
NEGATIVE ODOUR NOTICEABLE?						

COLONY NAME	WEATHER CONDITIONS
DATE	🌡 ___ ☀ ⛅ ☁ 🌧 ❄
TIME	🚩 ___ ☐ ☐ ☐ ☐ ☐

INSPECTION

HIVE NUMBER	①	②	③	④	⑤	⑥

PRODUCTIVITY & REPRODUCTION

AMOUNT OF HONEY						
GENERAL POPULATION						
AMOUNT OF BROOD						
AMOUNT OF SPACE						

BEHAVIOUR & ACTIVITIES

USUAL ENTERING AND EXITING ACTIVITY?						
CALM BEHAVIOUR WHEN OPENING HIVE?						
BEES BRINGING POLLEN INTO HIVE?						
SIGNS OF ROBBERY AMONG THE BEES?						

HEALTH STATUS

BEES SEEM WEAK OR LAZY?						
HIGH AMOUNT OF DEAD BEES?						
QUEEN BEE IS PRESENT / IDENTIFIABLE?						
INFESTATION BY ANTS / ANTS PRESENT?						
INFESTATION BY WAX MOTH / WAX MOTH PRESENT?						
NEGATIVE ODOUR NOTICEABLE?						

COLONY NAME	WEATHER CONDITIONS

🗓 DATE	🌡 _____ ☀ ⛅ 🌧 ⛈ ❄
🕐 TIME	🚩 _____ ☐ ☐ ☐ ☐ ☐

INSPECTION

HIVE NUMBER	①	②	③	④	⑤	⑥

PRODUCTIVITY & REPRODUCTION

AMOUNT OF HONEY						
GENERAL POPULATION						
AMOUNT OF BROOD						
AMOUNT OF SPACE						

BEHAVIOUR & ACTIVITIES

USUAL ENTERING AND EXITING ACTIVITY?						
CALM BEHAVIOUR WHEN OPENING HIVE?						
BEES BRINGING POLLEN INTO HIVE?						
SIGNS OF ROBBERY AMONG THE BEES?						

HEALTH STATUS

BEES SEEM WEAK OR LAZY?						
HIGH AMOUNT OF DEAD BEES?						
QUEEN BEE IS PRESENT / IDENTIFIABLE?						
INFESTATION BY ANTS / ANTS PRESENT?						
INFESTATION BY WAX MOTH / WAX MOTH PRESENT?						
NEGATIVE ODOUR NOTICEABLE?						

COLONY NAME	WEATHER CONDITIONS

🏠 COLONY NAME
📅 DATE
🕐 TIME

WEATHER CONDITIONS

🌡 ____ ☀️ ⛅ 🌧 ⛈ ❄️

🚩 ____ ☐ ☐ ☐ ☐ ☐

INSPECTION

	1	2	3	4	5	6
🐝 HIVE NUMBER	①	②	③	④	⑤	⑥

PRODUCTIVITY & REPRODUCTION

🍯 AMOUNT OF HONEY						
🐝 GENERAL POPULATION						
🐝 AMOUNT OF BROOD						
⬡ AMOUNT OF SPACE						

BEHAVIOUR & ACTIVITIES

USUAL ENTERING AND EXITING ACTIVITY?						
CALM BEHAVIOUR WHEN OPENING HIVE?						
BEES BRINGING POLLEN INTO HIVE?						
SIGNS OF ROBBERY AMONG THE BEES?						

HEALTH STATUS

BEES SEEM WEAK OR LAZY?						
HIGH AMOUNT OF DEAD BEES?						
QUEEN BEE IS PRESENT / IDENTIFIABLE?						
INFESTATION BY ANTS / ANTS PRESENT?						
INFESTATION BY WAX MOTH / WAX MOTH PRESENT?						
NEGATIVE ODOUR NOTICEABLE?						

COLONY NAME	
DATE	
TIME	

WEATHER CONDITIONS

Temperature — ☀ ⛅ 🌧 ⛈ ❄

Wind — ☐ ☐ ☐ ☐ ☐

INSPECTION

HIVE NUMBER	1	2	3	4	5	6

PRODUCTIVITY & REPRODUCTION

AMOUNT OF HONEY						
GENERAL POPULATION						
AMOUNT OF BROOD						
AMOUNT OF SPACE						

BEHAVIOUR & ACTIVITIES

USUAL ENTERING AND EXITING ACTIVITY?						
CALM BEHAVIOUR WHEN OPENING HIVE?						
BEES BRINGING POLLEN INTO HIVE?						
SIGNS OF ROBBERY AMONG THE BEES?						

HEALTH STATUS

BEES SEEM WEAK OR LAZY?						
HIGH AMOUNT OF DEAD BEES?						
QUEEN BEE IS PRESENT / IDENTIFIABLE?						
INFESTATION BY ANTS / ANTS PRESENT?						
INFESTATION BY WAX MOTH / WAX MOTH PRESENT?						
NEGATIVE ODOUR NOTICEABLE?						

COLONY NAME

DATE

TIME

WEATHER CONDITIONS

	☀	⛅	☁	🌧	❄
	☐	☐	☐	☐	☐

INSPECTION

HIVE NUMBER	①	②	③	④	⑤	⑥

PRODUCTIVITY & REPRODUCTION

AMOUNT OF HONEY						
GENERAL POPULATION						
AMOUNT OF BROOD						
AMOUNT OF SPACE						

BEHAVIOUR & ACTIVITIES

USUAL ENTERING AND EXITING ACTIVITY?						
CALM BEHAVIOUR WHEN OPENING HIVE?						
BEES BRINGING POLLEN INTO HIVE?						
SIGNS OF ROBBERY AMONG THE BEES?						

HEALTH STATUS

BEES SEEM WEAK OR LAZY?						
HIGH AMOUNT OF DEAD BEES?						
QUEEN BEE IS PRESENT / IDENTIFIABLE?						
INFESTATION BY ANTS / ANTS PRESENT?						
INFESTATION BY WAX MOTH / WAX MOTH PRESENT?						
NEGATIVE ODOUR NOTICEABLE?						

COLONY NAME		WEATHER CONDITIONS					
DATE		🌡 ——	☀	⛅	🌧	⛈	❄
TIME		🚩 ——	☐	☐	☐	☐	☐

INSPECTION

HIVE NUMBER	①	②	③	④	⑤	⑥

PRODUCTIVITY & REPRODUCTION

AMOUNT OF HONEY						
GENERAL POPULATION						
AMOUNT OF BROOD						
AMOUNT OF SPACE						

BEHAVIOUR & ACTIVITIES

USUAL ENTERING AND EXITING ACTIVITY?						
CALM BEHAVIOUR WHEN OPENING HIVE?						
BEES BRINGING POLLEN INTO HIVE?						
SIGNS OF ROBBERY AMONG THE BEES?						

HEALTH STATUS

BEES SEEM WEAK OR LAZY?						
HIGH AMOUNT OF DEAD BEES?						
QUEEN BEE IS PRESENT / IDENTIFIABLE?						
INFESTATION BY ANTS / ANTS PRESENT?						
INFESTATION BY WAX MOTH / WAX MOTH PRESENT?						
NEGATIVE ODOUR NOTICEABLE?						

COLONY NAME

DATE

TIME

WEATHER CONDITIONS

🌡 ____   ☀ ⛅ ☁ 🌧 ❄

🚩 ____   ☐ ☐ ☐ ☐ ☐

INSPECTION

HIVE NUMBER	①	②	③	④	⑤	⑥

PRODUCTIVITY & REPRODUCTION

AMOUNT OF HONEY						
GENERAL POPULATION						
AMOUNT OF BROOD						
AMOUNT OF SPACE						

BEHAVIOUR & ACTIVITIES

USUAL ENTERING AND EXITING ACTIVITY?						
CALM BEHAVIOUR WHEN OPENING HIVE?						
BEES BRINGING POLLEN INTO HIVE?						
SIGNS OF ROBBERY AMONG THE BEES?						

HEALTH STATUS

BEES SEEM WEAK OR LAZY?						
HIGH AMOUNT OF DEAD BEES?						
QUEEN BEE IS PRESENT / IDENTIFIABLE?						
INFESTATION BY ANTS / ANTS PRESENT?						
INFESTATION BY WAX MOTH / WAX MOTH PRESENT?						
NEGATIVE ODOUR NOTICEABLE?						

COLONY NAME		WEATHER CONDITIONS

🏠 COLONY NAME
📅 DATE
🕐 TIME

WEATHER CONDITIONS

🌡️ _____ ☀️ ⛅ 🌧️ ⛈️ ❄️

🚩 _____ ☐ ☐ ☐ ☐ ☐

INSPECTION

🐝 HIVE NUMBER	①	②	③	④	⑤	⑥

PRODUCTIVITY & REPRODUCTION

AMOUNT OF HONEY						
GENERAL POPULATION						
AMOUNT OF BROOD						
AMOUNT OF SPACE						

BEHAVIOUR & ACTIVITIES

USUAL ENTERING AND EXITING ACTIVITY?						
CALM BEHAVIOUR WHEN OPENING HIVE?						
BEES BRINGING POLLEN INTO HIVE?						
SIGNS OF ROBBERY AMONG THE BEES?						

HEALTH STATUS

BEES SEEM WEAK OR LAZY?						
HIGH AMOUNT OF DEAD BEES?						
QUEEN BEE IS PRESENT / IDENTIFIABLE?						
INFESTATION BY ANTS / ANTS PRESENT?						
INFESTATION BY WAX MOTH / WAX MOTH PRESENT?						
NEGATIVE ODOUR NOTICEABLE?						

COLONY NAME		WEATHER CONDITIONS					
DATE		🌡 —	☀	⛅	☁	🌧	❄
TIME		🚩 —	☐	☐	☐	☐	☐

INSPECTION

HIVE NUMBER	1	2	3	4	5	6

PRODUCTIVITY & REPRODUCTION

AMOUNT OF HONEY						
GENERAL POPULATION						
AMOUNT OF BROOD						
AMOUNT OF SPACE						

BEHAVIOUR & ACTIVITIES

USUAL ENTERING AND EXITING ACTIVITY?						
CALM BEHAVIOUR WHEN OPENING HIVE?						
BEES BRINGING POLLEN INTO HIVE?						
SIGNS OF ROBBERY AMONG THE BEES?						

HEALTH STATUS

BEES SEEM WEAK OR LAZY?						
HIGH AMOUNT OF DEAD BEES?						
QUEEN BEE IS PRESENT / IDENTIFIABLE?						
INFESTATION BY ANTS / ANTS PRESENT?						
INFESTATION BY WAX MOTH / WAX MOTH PRESENT?						
NEGATIVE ODOUR NOTICEABLE?						

COLONY NAME		WEATHER CONDITIONS					
DATE		🌡 ___	☀	⛅	☁	🌧	❄
TIME		🚩 ___	☐	☐	☐	☐	☐

INSPECTION

HIVE NUMBER	①	②	③	④	⑤	⑥

PRODUCTIVITY & REPRODUCTION

AMOUNT OF HONEY						
GENERAL POPULATION						
AMOUNT OF BROOD						
AMOUNT OF SPACE						

BEHAVIOUR & ACTIVITIES

USUAL ENTERING AND EXITING ACTIVITY?						
CALM BEHAVIOUR WHEN OPENING HIVE?						
BEES BRINGING POLLEN INTO HIVE?						
SIGNS OF ROBBERY AMONG THE BEES?						

HEALTH STATUS

BEES SEEM WEAK OR LAZY?						
HIGH AMOUNT OF DEAD BEES?						
QUEEN BEE IS PRESENT / IDENTIFIABLE?						
INFESTATION BY ANTS / ANTS PRESENT?						
INFESTATION BY WAX MOTH / WAX MOTH PRESENT?						
NEGATIVE ODOUR NOTICEABLE?						

	COLONY NAME	
	DATE	
	TIME	

WEATHER CONDITIONS

🌡 ___ ☀ ⛅ 🌧 ⛈ ❄

🚩 ___ ☐ ☐ ☐ ☐ ☐

INSPECTION

HIVE NUMBER	①	②	③	④	⑤	⑥

PRODUCTIVITY & REPRODUCTION

AMOUNT OF HONEY						
GENERAL POPULATION						
AMOUNT OF BROOD						
AMOUNT OF SPACE						

BEHAVIOUR & ACTIVITIES

USUAL ENTERING AND EXITING ACTIVITY?						
CALM BEHAVIOUR WHEN OPENING HIVE?						
BEES BRINGING POLLEN INTO HIVE?						
SIGNS OF ROBBERY AMONG THE BEES?						

HEALTH STATUS

BEES SEEM WEAK OR LAZY?						
HIGH AMOUNT OF DEAD BEES?						
QUEEN BEE IS PRESENT / IDENTIFIABLE?						
INFESTATION BY ANTS / ANTS PRESENT?						
INFESTATION BY WAX MOTH / WAX MOTH PRESENT?						
NEGATIVE ODOUR NOTICEABLE?						

COLONY NAME		WEATHER CONDITIONS					
DATE		☼ ⛅ 🌧 ⛈ ❄					
TIME		▯ ▯ ▯ ▯ ▯					

INSPECTION

HIVE NUMBER	①	②	③	④	⑤	⑥

PRODUCTIVITY & REPRODUCTION

AMOUNT OF HONEY						
GENERAL POPULATION						
AMOUNT OF BROOD						
AMOUNT OF SPACE						

BEHAVIOUR & ACTIVITIES

USUAL ENTERING AND EXITING ACTIVITY?						
CALM BEHAVIOUR WHEN OPENING HIVE?						
BEES BRINGING POLLEN INTO HIVE?						
SIGNS OF ROBBERY AMONG THE BEES?						

HEALTH STATUS

BEES SEEM WEAK OR LAZY?						
HIGH AMOUNT OF DEAD BEES?						
QUEEN BEE IS PRESENT / IDENTIFIABLE?						
INFESTATION BY ANTS / ANTS PRESENT?						
INFESTATION BY WAX MOTH / WAX MOTH PRESENT?						
NEGATIVE ODOUR NOTICEABLE?						

COLONY NAME		WEATHER CONDITIONS					
DATE		🌡 ___	☀	⛅	☁	🌧	❄
TIME		🚩 ___	☐	☐	☐	☐	☐

INSPECTION

HIVE NUMBER	①	②	③	④	⑤	⑥

PRODUCTIVITY & REPRODUCTION

AMOUNT OF HONEY						
GENERAL POPULATION						
AMOUNT OF BROOD						
AMOUNT OF SPACE						

BEHAVIOUR & ACTIVITIES

USUAL ENTERING AND EXITING ACTIVITY?						
CALM BEHAVIOUR WHEN OPENING HIVE?						
BEES BRINGING POLLEN INTO HIVE?						
SIGNS OF ROBBERY AMONG THE BEES?						

HEALTH STATUS

BEES SEEM WEAK OR LAZY?						
HIGH AMOUNT OF DEAD BEES?						
QUEEN BEE IS PRESENT / IDENTIFIABLE?						
INFESTATION BY ANTS / ANTS PRESENT?						
INFESTATION BY WAX MOTH / WAX MOTH PRESENT?						
NEGATIVE ODOUR NOTICEABLE?						

COLONY NAME

DATE

TIME

WEATHER CONDITIONS

🌡	—	☀	⛅	☁	🌧	❄
🚩	—	☐	☐	☐	☐	☐

INSPECTION

HIVE NUMBER	①	②	③	④	⑤	⑥

PRODUCTIVITY & REPRODUCTION

AMOUNT OF HONEY						
GENERAL POPULATION						
AMOUNT OF BROOD						
AMOUNT OF SPACE						

BEHAVIOUR & ACTIVITIES

USUAL ENTERING AND EXITING ACTIVITY?						
CALM BEHAVIOUR WHEN OPENING HIVE?						
BEES BRINGING POLLEN INTO HIVE?						
SIGNS OF ROBBERY AMONG THE BEES?						

HEALTH STATUS

BEES SEEM WEAK OR LAZY?						
HIGH AMOUNT OF DEAD BEES?						
QUEEN BEE IS PRESENT / IDENTIFIABLE?						
INFESTATION BY ANTS / ANTS PRESENT?						
INFESTATION BY WAX MOTH / WAX MOTH PRESENT?						
NEGATIVE ODOUR NOTICEABLE?						

COLONY NAME		WEATHER CONDITIONS

🏠 COLONY NAME	
📅 DATE	
🕐 TIME	

WEATHER CONDITIONS

🌡 _____ ☀️ ⛅ 🌧 ⛈ ❄️

🚩 _____ ☐ ☐ ☐ ☐ ☐

INSPECTION

🐝 HIVE NUMBER	①	②	③	④	⑤	⑥

PRODUCTIVITY & REPRODUCTION

🍯 AMOUNT OF HONEY						
🐝 GENERAL POPULATION						
🐝 AMOUNT OF BROOD						
🐝 AMOUNT OF SPACE						

BEHAVIOUR & ACTIVITIES

🐝 USUAL ENTERING AND EXITING ACTIVITY?						
📦 CALM BEHAVIOUR WHEN OPENING HIVE?						
🐝 BEES BRINGING POLLEN INTO HIVE?						
🐝 SIGNS OF ROBBERY AMONG THE BEES?						

HEALTH STATUS

🐝 BEES SEEM WEAK OR LAZY?						
☠️ HIGH AMOUNT OF DEAD BEES?						
🐝 QUEEN BEE IS PRESENT / IDENTIFIABLE?						
🐜 INFESTATION BY ANTS / ANTS PRESENT?						
🦋 INFESTATION BY WAX MOTH / WAX MOTH PRESENT?						
👃 NEGATIVE ODOUR NOTICEABLE?						

COLONY NAME		WEATHER CONDITIONS					
DATE							
TIME							

INSPECTION

HIVE NUMBER	①	②	③	④	⑤	⑥

PRODUCTIVITY & REPRODUCTION

AMOUNT OF HONEY						
GENERAL POPULATION						
AMOUNT OF BROOD						
AMOUNT OF SPACE						

BEHAVIOUR & ACTIVITIES

USUAL ENTERING AND EXITING ACTIVITY?						
CALM BEHAVIOUR WHEN OPENING HIVE?						
BEES BRINGING POLLEN INTO HIVE?						
SIGNS OF ROBBERY AMONG THE BEES?						

HEALTH STATUS

BEES SEEM WEAK OR LAZY?						
HIGH AMOUNT OF DEAD BEES?						
QUEEN BEE IS PRESENT / IDENTIFIABLE?						
INFESTATION BY ANTS / ANTS PRESENT?						
INFESTATION BY WAX MOTH / WAX MOTH PRESENT?						
NEGATIVE ODOUR NOTICEABLE?						

COLONY NAME		WEATHER CONDITIONS					
DATE		🌡 ___	☀	⛅	🌧	⛈	❄
TIME		📶 ___	☐	☐	☐	☐	☐

INSPECTION

HIVE NUMBER	1	2	3	4	5	6

PRODUCTIVITY & REPRODUCTION

AMOUNT OF HONEY						
GENERAL POPULATION						
AMOUNT OF BROOD						
AMOUNT OF SPACE						

BEHAVIOUR & ACTIVITIES

USUAL ENTERING AND EXITING ACTIVITY?						
CALM BEHAVIOUR WHEN OPENING HIVE?						
BEES BRINGING POLLEN INTO HIVE?						
SIGNS OF ROBBERY AMONG THE BEES?						

HEALTH STATUS

BEES SEEM WEAK OR LAZY?						
HIGH AMOUNT OF DEAD BEES?						
QUEEN BEE IS PRESENT / IDENTIFIABLE?						
INFESTATION BY ANTS / ANTS PRESENT?						
INFESTATION BY WAX MOTH / WAX MOTH PRESENT?						
NEGATIVE ODOUR NOTICEABLE?						

COLONY NAME		WEATHER CONDITIONS					
DATE		🌡 ___	☀	⛅	🌧	⛈	❄
TIME		🚩 ___	☐	☐	☐	☐	☐

INSPECTION

HIVE NUMBER	(1)	(2)	(3)	(4)	(5)	(6)

PRODUCTIVITY & REPRODUCTION

AMOUNT OF HONEY						
GENERAL POPULATION						
AMOUNT OF BROOD						
AMOUNT OF SPACE						

BEHAVIOUR & ACTIVITIES

USUAL ENTERING AND EXITING ACTIVITY?						
CALM BEHAVIOUR WHEN OPENING HIVE?						
BEES BRINGING POLLEN INTO HIVE?						
SIGNS OF ROBBERY AMONG THE BEES?						

HEALTH STATUS

BEES SEEM WEAK OR LAZY?						
HIGH AMOUNT OF DEAD BEES?						
QUEEN BEE IS PRESENT / IDENTIFIABLE?						
INFESTATION BY ANTS / ANTS PRESENT?						
INFESTATION BY WAX MOTH / WAX MOTH PRESENT?						
NEGATIVE ODOUR NOTICEABLE?						

	COLONY NAME
	DATE
	TIME

WEATHER CONDITIONS

		☀	⛅	☁	🌧	❄
🌡	—					
🎏	—	☐	☐	☐	☐	☐

INSPECTION

HIVE NUMBER	①	②	③	④	⑤	⑥

PRODUCTIVITY & REPRODUCTION

AMOUNT OF HONEY						
GENERAL POPULATION						
AMOUNT OF BROOD						
AMOUNT OF SPACE						

BEHAVIOUR & ACTIVITIES

USUAL ENTERING AND EXITING ACTIVITY?					
CALM BEHAVIOUR WHEN OPENING HIVE?					
BEES BRINGING POLLEN INTO HIVE?					
SIGNS OF ROBBERY AMONG THE BEES?					

HEALTH STATUS

BEES SEEM WEAK OR LAZY?					
HIGH AMOUNT OF DEAD BEES?					
QUEEN BEE IS PRESENT / IDENTIFIABLE?					
INFESTATION BY ANTS / ANTS PRESENT?					
INFESTATION BY WAX MOTH / WAX MOTH PRESENT?					
NEGATIVE ODOUR NOTICEABLE?					

COLONY NAME	WEATHER CONDITIONS

COLONY NAME
DATE
TIME

WEATHER CONDITIONS

🌡️ —— ☀️ ⛅ 🌧️ ⛈️ ❄️

🎏 —— ☐ ☐ ☐ ☐ ☐

INSPECTION

HIVE NUMBER	①	②	③	④	⑤	⑥

PRODUCTIVITY & REPRODUCTION

AMOUNT OF HONEY						
GENERAL POPULATION						
AMOUNT OF BROOD						
AMOUNT OF SPACE						

BEHAVIOUR & ACTIVITIES

USUAL ENTERING AND EXITING ACTIVITY?						
CALM BEHAVIOUR WHEN OPENING HIVE?						
BEES BRINGING POLLEN INTO HIVE?						
SIGNS OF ROBBERY AMONG THE BEES?						

HEALTH STATUS

BEES SEEM WEAK OR LAZY?						
HIGH AMOUNT OF DEAD BEES?						
QUEEN BEE IS PRESENT / IDENTIFIABLE?						
INFESTATION BY ANTS / ANTS PRESENT?						
INFESTATION BY WAX MOTH / WAX MOTH PRESENT?						
NEGATIVE ODOUR NOTICEABLE?						

COLONY NAME	
DATE	
TIME	

WEATHER CONDITIONS

🌡️ ____	☀️	⛅	🌧️	⛈️	❄️
🚩 ____	☐	☐	☐	☐	☐

INSPECTION

HIVE NUMBER	①	②	③	④	⑤	⑥

PRODUCTIVITY & REPRODUCTION

AMOUNT OF HONEY						
GENERAL POPULATION						
AMOUNT OF BROOD						
AMOUNT OF SPACE						

BEHAVIOUR & ACTIVITIES

USUAL ENTERING AND EXITING ACTIVITY?						
CALM BEHAVIOUR WHEN OPENING HIVE?						
BEES BRINGING POLLEN INTO HIVE?						
SIGNS OF ROBBERY AMONG THE BEES?						

HEALTH STATUS

BEES SEEM WEAK OR LAZY?						
HIGH AMOUNT OF DEAD BEES?						
QUEEN BEE IS PRESENT / IDENTIFIABLE?						
INFESTATION BY ANTS / ANTS PRESENT?						
INFESTATION BY WAX MOTH / WAX MOTH PRESENT?						
NEGATIVE ODOUR NOTICEABLE?						

COLONY NAME		WEATHER CONDITIONS					

DATE	
TIME	

INSPECTION

HIVE NUMBER	1	2	3	4	5	6

PRODUCTIVITY & REPRODUCTION

	1	2	3	4	5	6
AMOUNT OF HONEY						
GENERAL POPULATION						
AMOUNT OF BROOD						
AMOUNT OF SPACE						

BEHAVIOUR & ACTIVITIES

	1	2	3	4	5	6
USUAL ENTERING AND EXITING ACTIVITY?						
CALM BEHAVIOUR WHEN OPENING HIVE?						
BEES BRINGING POLLEN INTO HIVE?						
SIGNS OF ROBBERY AMONG THE BEES?						

HEALTH STATUS

	1	2	3	4	5	6
BEES SEEM WEAK OR LAZY?						
HIGH AMOUNT OF DEAD BEES?						
QUEEN BEE IS PRESENT / IDENTIFIABLE?						
INFESTATION BY ANTS / ANTS PRESENT?						
INFESTATION BY WAX MOTH / WAX MOTH PRESENT?						
NEGATIVE ODOUR NOTICEABLE?						

COLONY NAME

DATE

TIME

WEATHER CONDITIONS

🌡 ——	☀	⛅	🌧	⛈	❄
🚩 ——	☐	☐	☐	☐	☐

INSPECTION

HIVE NUMBER	1	2	3	4	5	6

PRODUCTIVITY & REPRODUCTION

AMOUNT OF HONEY						
GENERAL POPULATION						
AMOUNT OF BROOD						
AMOUNT OF SPACE						

BEHAVIOUR & ACTIVITIES

USUAL ENTERING AND EXITING ACTIVITY?						
CALM BEHAVIOUR WHEN OPENING HIVE?						
BEES BRINGING POLLEN INTO HIVE?						
SIGNS OF ROBBERY AMONG THE BEES?						

HEALTH STATUS

BEES SEEM WEAK OR LAZY?						
HIGH AMOUNT OF DEAD BEES?						
QUEEN BEE IS PRESENT / IDENTIFIABLE?						
INFESTATION BY ANTS / ANTS PRESENT?						
INFESTATION BY WAX MOTH / WAX MOTH PRESENT?						
NEGATIVE ODOUR NOTICEABLE?						

COLONY NAME		WEATHER CONDITIONS

🏠 **COLONY NAME**		**WEATHER CONDITIONS**
📅 **DATE**		🌡 ___ ☀ ⛅ 🌧 ⛈ ❄
🕐 **TIME**		🚩 ___ ☐ ☐ ☐ ☐ ☐

INSPECTION

🐝 HIVE NUMBER	①	②	③	④	⑤	⑥

PRODUCTIVITY & REPRODUCTION

🍯 AMOUNT OF HONEY						
🐝 GENERAL POPULATION						
🐝 AMOUNT OF BROOD						
🍯 AMOUNT OF SPACE						

BEHAVIOUR & ACTIVITIES

🐝 USUAL ENTERING AND EXITING ACTIVITY?						
📦 CALM BEHAVIOUR WHEN OPENING HIVE?						
🐝 BEES BRINGING POLLEN INTO HIVE?						
🐝 SIGNS OF ROBBERY AMONG THE BEES?						

HEALTH STATUS

🐝 BEES SEEM WEAK OR LAZY?						
☠ HIGH AMOUNT OF DEAD BEES?						
🐝 QUEEN BEE IS PRESENT / IDENTIFIABLE?						
🐜 INFESTATION BY ANTS / ANTS PRESENT?						
🦋 INFESTATION BY WAX MOTH / WAX MOTH PRESENT?						
👃 NEGATIVE ODOUR NOTICEABLE?						

COLONY NAME		WEATHER CONDITIONS

COLONY NAME
DATE
TIME

WEATHER CONDITIONS

🌡 ___ ☀ ⛅ ☁ 🌧 ❄

🚩 ___ ☐ ☐ ☐ ☐ ☐

INSPECTION						
HIVE NUMBER	1	2	3	4	5	6

PRODUCTIVITY & REPRODUCTION						
AMOUNT OF HONEY						
GENERAL POPULATION						
AMOUNT OF BROOD						
AMOUNT OF SPACE						

BEHAVIOUR & ACTIVITIES						
USUAL ENTERING AND EXITING ACTIVITY?						
CALM BEHAVIOUR WHEN OPENING HIVE?						
BEES BRINGING POLLEN INTO HIVE?						
SIGNS OF ROBBERY AMONG THE BEES?						

HEALTH STATUS						
BEES SEEM WEAK OR LAZY?						
HIGH AMOUNT OF DEAD BEES?						
QUEEN BEE IS PRESENT / IDENTIFIABLE?						
INFESTATION BY ANTS / ANTS PRESENT?						
INFESTATION BY WAX MOTH / WAX MOTH PRESENT?						
NEGATIVE ODOUR NOTICEABLE?						

COLONY NAME		WEATHER CONDITIONS					

COLONY NAME

🏠 **COLONY NAME**

📅 **DATE**

🕐 **TIME**

WEATHER CONDITIONS

🌡 ——   ☀ ⛅ 🌧 ⛈ ❄

🚩 ——   ☐ ☐ ☐ ☐ ☐

INSPECTION

	HIVE NUMBER	①	②	③	④	⑤	⑥

PRODUCTIVITY & REPRODUCTION

AMOUNT OF HONEY						
GENERAL POPULATION						
AMOUNT OF BROOD						
AMOUNT OF SPACE						

BEHAVIOUR & ACTIVITIES

USUAL ENTERING AND EXITING ACTIVITY?						
CALM BEHAVIOUR WHEN OPENING HIVE?						
BEES BRINGING POLLEN INTO HIVE?						
SIGNS OF ROBBERY AMONG THE BEES?						

HEALTH STATUS

BEES SEEM WEAK OR LAZY?						
HIGH AMOUNT OF DEAD BEES?						
QUEEN BEE IS PRESENT / IDENTIFIABLE?						
INFESTATION BY ANTS / ANTS PRESENT?						
INFESTATION BY WAX MOTH / WAX MOTH PRESENT?						
NEGATIVE ODOUR NOTICEABLE?						

	COLONY NAME	
	DATE	
	TIME	

WEATHER CONDITIONS

🌡 ____ ☀ ⛅ 🌧 ⛈ ❄

🚩 ____ ☐ ☐ ☐ ☐ ☐

INSPECTION

	HIVE NUMBER	①	②	③	④	⑤	⑥

PRODUCTIVITY & REPRODUCTION

	AMOUNT OF HONEY						
	GENERAL POPULATION						
	AMOUNT OF BROOD						
	AMOUNT OF SPACE						

BEHAVIOUR & ACTIVITIES

	USUAL ENTERING AND EXITING ACTIVITY?						
	CALM BEHAVIOUR WHEN OPENING HIVE?						
	BEES BRINGING POLLEN INTO HIVE?						
	SIGNS OF ROBBERY AMONG THE BEES?						

HEALTH STATUS

	BEES SEEM WEAK OR LAZY?						
	HIGH AMOUNT OF DEAD BEES?						
	QUEEN BEE IS PRESENT / IDENTIFIABLE?						
	INFESTATION BY ANTS / ANTS PRESENT?						
	INFESTATION BY WAX MOTH / WAX MOTH PRESENT?						
	NEGATIVE ODOUR NOTICEABLE?						

COLONY NAME	WEATHER CONDITIONS
DATE	🌡 — ☀ ⛅ ☁ 🌧 ❄
TIME	🚩 — ☐ ☐ ☐ ☐ ☐

INSPECTION

HIVE NUMBER	①	②	③	④	⑤	⑥

PRODUCTIVITY & REPRODUCTION

AMOUNT OF HONEY						
GENERAL POPULATION						
AMOUNT OF BROOD						
AMOUNT OF SPACE						

BEHAVIOUR & ACTIVITIES

USUAL ENTERING AND EXITING ACTIVITY?						
CALM BEHAVIOUR WHEN OPENING HIVE?						
BEES BRINGING POLLEN INTO HIVE?						
SIGNS OF ROBBERY AMONG THE BEES?						

HEALTH STATUS

BEES SEEM WEAK OR LAZY?						
HIGH AMOUNT OF DEAD BEES?						
QUEEN BEE IS PRESENT / IDENTIFIABLE?						
INFESTATION BY ANTS / ANTS PRESENT?						
INFESTATION BY WAX MOTH / WAX MOTH PRESENT?						
NEGATIVE ODOUR NOTICEABLE?						

COLONY NAME	WEATHER CONDITIONS
DATE	🌡 ____ ☀ ⛅ 🌧 ⛈ ❄
TIME	🚩 ____ ☐ ☐ ☐ ☐ ☐

INSPECTION

HIVE NUMBER	①	②	③	④	⑤	⑥

PRODUCTIVITY & REPRODUCTION

AMOUNT OF HONEY						
GENERAL POPULATION						
AMOUNT OF BROOD						
AMOUNT OF SPACE						

BEHAVIOUR & ACTIVITIES

USUAL ENTERING AND EXITING ACTIVITY?						
CALM BEHAVIOUR WHEN OPENING HIVE?						
BEES BRINGING POLLEN INTO HIVE?						
SIGNS OF ROBBERY AMONG THE BEES?						

HEALTH STATUS

BEES SEEM WEAK OR LAZY?						
HIGH AMOUNT OF DEAD BEES?						
QUEEN BEE IS PRESENT / IDENTIFIABLE?						
INFESTATION BY ANTS / ANTS PRESENT?						
INFESTATION BY WAX MOTH / WAX MOTH PRESENT?						
NEGATIVE ODOUR NOTICEABLE?						

	COLONY NAME		WEATHER CONDITIONS					
	DATE			☀	⛅	🌧	⛈	❄
	TIME			☐	☐	☐	☐	☐

INSPECTION

HIVE NUMBER		①	②	③	④	⑤	⑥

PRODUCTIVITY & REPRODUCTION

AMOUNT OF HONEY						
GENERAL POPULATION						
AMOUNT OF BROOD						
AMOUNT OF SPACE						

BEHAVIOUR & ACTIVITIES

USUAL ENTERING AND EXITING ACTIVITY?						
CALM BEHAVIOUR WHEN OPENING HIVE?						
BEES BRINGING POLLEN INTO HIVE?						
SIGNS OF ROBBERY AMONG THE BEES?						

HEALTH STATUS

BEES SEEM WEAK OR LAZY?						
HIGH AMOUNT OF DEAD BEES?						
QUEEN BEE IS PRESENT / IDENTIFIABLE?						
INFESTATION BY ANTS / ANTS PRESENT?						
INFESTATION BY WAX MOTH / WAX MOTH PRESENT?						
NEGATIVE ODOUR NOTICEABLE?						

COLONY NAME	WEATHER CONDITIONS

COLONY NAME

DATE

TIME

WEATHER CONDITIONS

🌡 ____   ☀ ⛅ ☁ 🌧 ❄

🏳 ____   ☐ ☐ ☐ ☐ ☐

INSPECTION

HIVE NUMBER	1	2	3	4	5	6

PRODUCTIVITY & REPRODUCTION

AMOUNT OF HONEY						
GENERAL POPULATION						
AMOUNT OF BROOD						
AMOUNT OF SPACE						

BEHAVIOUR & ACTIVITIES

USUAL ENTERING AND EXITING ACTIVITY?						
CALM BEHAVIOUR WHEN OPENING HIVE?						
BEES BRINGING POLLEN INTO HIVE?						
SIGNS OF ROBBERY AMONG THE BEES?						

HEALTH STATUS

BEES SEEM WEAK OR LAZY?						
HIGH AMOUNT OF DEAD BEES?						
QUEEN BEE IS PRESENT / IDENTIFIABLE?						
INFESTATION BY ANTS / ANTS PRESENT?						
INFESTATION BY WAX MOTH / WAX MOTH PRESENT?						
NEGATIVE ODOUR NOTICEABLE?						

Made in the USA
Las Vegas, NV
29 November 2024

12911663R00056